IMPLEMENTIN(
INNOVATIONS

M000289746
EDUCATION, AND PRACTICE

IMPLEMENTING BIOMEDICAL INNOVATIONS INTO HEALTH, EDUCATION, AND PRACTICE

PREPARING TOMORROW'S PHYSICIANS

JAMES O. WOOLLISCROFT, MD, MACP, FRCP

Lyle C. Roll Professor of Medicine
Professor of Internal Medicine and Learning Health Sciences
University of Michigan Medical School
Ann Arbor, MI, United States

ACADEMIC PRESS

An imprint of Elsevier

ELSEVIER

Academic Press is an imprint of Elsevier
125 London Wall, London EC2Y 5AS, United Kingdom
525 B Street, Suite 1650, San Diego, CA 92101, United States
50 Hampshire Street, 5th Floor, Cambridge, MA 02139, United States
The Boulevard, Langford Lane, Kidlington, Oxford OX5 1GB, United Kingdom

Library of Congress Cataloging-in-Publication Data
A catalog record for this book is available from the Library of Congress

British Library Cataloguing-in-Publication Data
A catalogue record for this book is available from the British Library

ISBN 978-0-12-819620-5

For information on all Academic Press publications
visit our website at https://www.elsevier.com/books-and-journals

Publisher: Stacy Masucci
Acquisition Editor: Tari Broderick
Editorial Project Manager: Megan Ashdown
Production Project Manager: Kiruthika Govindaraju
Cover Designer: Victoria Bornstein

Typeset by SPi Global, India

Working together
to grow libraries in
developing countries

www.elsevier.com • www.bookaid.org

Dedication

To my wife Elizabeth for her support, love and patience as she "listened" to many versions of this book being "voice to text written"; our children Michael, Matthew, Marc, and Elizabeth Kate; and our daughters-in-law Tracey, Aleyna, Zenka and grandchildren Benjamin, Jacob, Madelyn, and Abigail who bring joy and fulfillment. And to future generations of physicians who will assume the mantle of healer and professional.

Contents

I

Setting the stage

II

Constants in medicine

III

Disruptors

IV

Implications for educators

Preface

The scientific knowledge upon which modern medicine is based is expanding our understanding of health and disease at an ever accelerating pace. Coupled with computational and technologic advances, how will the practice of clinical medicine look in mid-21st century? What are the educational implications for the preparation of future physicians?

Through interviews and assessment of trends in the biomedical sciences and related technologies, a picture of what the future of clinical medicine will entail has emerged. Clinical medicine in midcentury will be profoundly altered. Our knowledge of the complex relationships among our genome, microbiome, behavior and environment will continue to grow and inform our understanding of health and the progression to disease. "Healthcare" will be a reality as individuals at risk for disease will be identified and interventions to maintain health will begin much earlier than they do today. Sensors will enable continuous monitoring of chronic conditions, biomarkers and physiologic parameters. Artificial intelligence and machine learning algorithms will interpret images, monitor and manage respiratory parameters for intubated patients, and aid in diagnosis. Regenerative medicine will transform the approach to traumatic injury and organ failure. The disruption in medical practice will be widespread.

Recognizing that technology and scientific advances will greatly impact medicine, fundamental aspects of clinical practice will surely endure. Medicine is a human to human interaction. Communication, trustworthiness, and ethical interactions are some of the immutable aspects of physicianhood that will continue.

The educational preparation of physicians from the time they enter college to the completion of their formal training is a 10+ year journey. Therefore, it behooves medical educators, and all interested in the preparation of our future physicians, to thoughtfully consider whether the educational and experiential foundation laid in premedical, medical, and graduate medical education curricula is preparing them for practice in midcentury or is simply a reflection of the current reality.

It is the goal of this book to survey the directionality and implications of scientific and technologic advances that will impact medicine, identify the aspects of physicianhood that are immutable, and explore the implications for education across the continuum from premedical requirements to graduate medical education. Just as it is envisioned that clinical practice will be disrupted, so too will the educational enterprise be disrupted.

Recognizing the wisdom in an oft repeated proverb "it's difficult to make predictions, especially about the future," the projections and the attendant recommendations presented in this book are offered with humility and the knowledge that there will be developments we cannot envision that will dramatically impact the practice of clinical medicine and will require the continued attention of medical educators. This is the destination for which our future professionals must be prepared. It is an exciting journey with great consequences for our students on the path to physicianhood, for their future patients, for our academic medical centers, and for society at large. This book is intended for all those students, trainees, and professionals in medicine and in research who are interested in shaping the future of health and clinical practice.

James O. Woolliscroft

Acknowledgments

I want to thank the many distinguished individuals who freely shared their perspectives as to what the future will bring. Their insight, vision, and drive are creating this future. I am grateful for their willingness to "give forward" in our mutual desire to enhance the preparation of future physicians and enhance the care these physicians will provide to millions.

The development and production of this work has involved many steps. I thank Cindy Ellis who made the travel and interview arrangements and transcribed the many hours of interview recordings; Victoria Bornstein who has creatively captured the essence of the future as well as the enduring aspects of medicine through her illustrations; and the team at Elsevier, Tari Broderick, Senior Editor, Megan Ashdown, Editorial Project Manager, and Kiruthika Govindaraju, Project Manager, for their proficiency and skill in moving from manuscript to published work.

A special thank you to Jasna Markovac, PhD. The journey from concept to completion of this book has been intellectually stimulating, challenging and sometimes laborious. Her encouragement to persevere, professional editing expertise, and adroit guidance through the publishing process made this work possible.

Acknowledgments

Introduction

Envisioning the future of medicine seems at best presumptuous. After all, those of us who have the privilege of long professional careers readily identify advances that we could not have imagined 30, 40 or more years ago. Advances that have radically changed the practice of clinical medicine. Imaging, biologics and whole new classes of therapeutic agents, biomarkers, the proliferation of massive healthcare systems, and the list goes on. Indeed, the one thing that we can say with certainty is that the practice of clinical medicine in mid-21st century will be very different than it is now. But how different and what might be the implications of envisioned changes?

Through interviews with nationally and internationally recognized leaders in information technology, biomedical sciences, life science entrepreneurs, clinical systems, engineering and more, an understanding has emerged of the directionality and pace of change that will impact medical practice. Evidence from the literature supporting their perspectives is presented. While not intended to be comprehensive it corroborates the legitimacy of their views. It is inevitable that several trends with considerable disruptive potential will continue, it is also likely that unforeseen major disruptors will occur.

So why attempt to discern what the future practice of clinical medicine will entail? To become a physician is a lengthy process that requires major investments in time and resources by both the individual and the institution. An advantage that medical educators have is that the desired end product of the educational process provides the goal to which all learning experiences should build. When disruptive changes are anticipated in the skills and knowledge necessary for clinical practice, a critical consideration of how best to lay the necessary educational foundation for a lifetime of professional practice is needed. Hence, it is imperative that there be a vision of the clinical practice of the future. But the importance of optimally preparing future physicians for the realities they will encounter extends beyond just our students and trainees. If the promises inherent in scientific and technologic advances are to become reality, it is necessary that these future physicians be skilled in understanding and, where appropriate, applying those advances for the betterment of their patients and society at large.

It is expected that the pace of understanding of the basis for health and disease will continue to accelerate. This will lead to new approaches to diagnosis and therapy. It will also result in new classifications of disease. These advances will have important implications for medical educators. But are there developments on the horizon that will require knowledge and

skills not traditionally considered to be integral to the education of physicians? As major technologic advances have led to fundamental changes in
our understanding of health and disease and augured dramatic changes in
the practice of medicine throughout the last 150 years, we are certainly at
the threshold of a similar, if not even larger, technology driven revolution.

"Medicine is heavily influenced by technology. So when technology was
growing, as most industrial things did, on a geometric kind of scale, it was
easy to keep up. Somebody discovers a stethoscope, somebody discovers
ether, these were milestones and manageable. You'd start off in medicine, and
yes every generation would have a different skill set, a little better skill set
than the last, but it worked. You could read about the new drugs and gave the
best medical treatment you could.

That's no longer the case. We're now at the dawn of things changing so
rapidly that it's not conceivable to even keep up within the professional space.
The impact that this will have upon the field (medicine) is mind boggling."

Edward Schulak

As a cohesive view of the practice of clinical medicine in the mid-
21st century has not yet emerged, what knowledge and skills should
be, (1) foundational, (2) important but not essential, and (3) no longer
necessary? The educational continuum from premedical to graduate
medical education must be rigorously evaluated and the contributions
of each component considered. The implications of the saying, "50%
of what we teach you in medical school will prove to be incorrect, we
just don't know what 50%," does not excuse us from the weighty responsibility of attempting to discern the future. This requires thoughtful discussion and planning. It is imperative that medical educators,
accrediting bodies, professional organizations and public stakeholders engage in this conversation. The fundamental question of how to
best prepare medical students and residents for a lifetime of practice,
given the foreseeable and unforeseeable changes, is of paramount importance not only to medical educators, students and trainees, but also
our nation and beyond.

This book explores developments that have the potential to radically disrupt and change medical practice as well as new skills and
aptitudes that will be necessary for the midcentury physician. Readers
should critically assess the validity of the evidence supporting the
perspectives as well as the educational implications thereof. The goal
of this book is not to be the definitive guide, but rather to engender
thoughtful conversations regarding how best to develop the foundational knowledge skills and aptitudes of future physicians and other
healthcare professionals.

"…watches that may be able to pre-determine whether you're within a week or two of a heart attack…contact lens that measure all of these things (biomarkers), …diabetic devices that monitor 24 hours a day, … these devices, they're all coming. … genetic diseases that folks might have to confront decades down the line and what they can do for preventative medicine to avoid them.

You obviously have privacy concerns that have to be addressed but it is going to be a whole different life, 5, 10, 30 years from now. It's breath-taking, it's going to be something else."

Representative Fred Upton

While we will consider the clinical and educational implications of scientific and technologic breakthroughs, it is important to ensure that the essence of what it means to be a physician remains an area of emphasis. There are immutable aspects of physicianhood that have been constant over centuries. These "constants" must remain central to educational programs. Arguably this is the most important role for medical educators and the Academy. We are at a juncture where the fundamental role of physicians as "healers" must be consciously embraced. Medicine is a human to human interaction. While appreciating the potential of our scientific and technologic advances, we must also recognize the inherent risks they entail and the social and ethical issues they raise.

SECTION I

SETTING
THE STAGE

Medical education is goal oriented. Its central purpose is to establish in future physicians the foundation for decades of the highest quality clinical practice applying the latest scientific understanding for the benefit of their patients in an ethical and humane manner. But this aspirational goal, absent reasonably well defined specific components is unlikely to be achieved. Hence, the focus on trying to divine the future of clinical medicine midcentury.

But beyond the difficulties in predicting future developments, the tendency for educators to focus on what is known and well established rather than what will be is common and oftentimes rate limiting. So let us try to imagine what that future might entail.

Not only is it useful to think about what the future might bring, it is also important to understand the journey that has brought the academy and the medical education enterprise to the present. Many of the decisions were based on visionary thinking and leadership. Others in reaction to external stimuli. Medical education and academic medical centers have evolved greatly over the last century and continue to evolve. It is incumbent upon medical educators to ensure that the aspirational goal continues to be central to the purpose of all academic medical centers. We have arrived at this juncture in the history of medicine and medical education not through happenstance but rather through thousands of decisions that collectively have defined our current reality. So too will the future reality of medicine be defined by our decisions and those of our successors.

Life as a mid-21st century physician

As all of us are "prisoners of our autobiography" it is oftentimes difficult to try to imagine a reality fundamentally different from our own. The scenarios presented here depict a hypothetical day in the life of mid-21st century physicians. The goal is to project a version of future reality that will facilitate our consideration of how scientific and technologic changes might alter medical practice and how each of these clinicians might best be prepared. These scenarios are presented with the additional expectation that the future practice of clinical medicine will also be influenced by developments that we cannot even imagine.

Implementing Biomedical Innovations into Health, Education, and Practice
https://doi.org/10.1016/B978-0-12-819620-5.00001-1

Context: The mid-21st century healthcare landscape

Financing

The United States has finally embraced a value-based payment system for health and medical care. The confluence of the aged baby boomer generation, the realization that longer lifespans also meant that individuals with chronic disease would need care for longer periods of time, the near bankruptcy of the Medicare Trust Fund, and the critical need to invest in other social goods finally led to this reform in the late 2020s.

One of the conundrums was how best to allocate responsibility for care and payment across the spectrum from "at risk" to "manifest disease." While it was recognized that many of the "highest value" interventions focused on prevention and restoration of health prior to the development of symptomatic disease, the fact that savings would not be realized for decades was a major impediment to widespread adoption. This long-standing problem in the past led to a marginalization of interventions to maintain health and prevent the development of disease. Now, a compromise was reached by putting individual states (rather than the US federal government) in charge of ensuring that nationally established basic preventive and health enhancing measures were implemented.

While a national single-payer system was seriously considered, the final agreement was to use nationally licensed insurance carriers and a multitier system. A baseline, or safety net, federally run system was developed for all citizens at or below 125% of the federally determined poverty level. Citizens choosing this coverage who were above that level paid premiums based on an income adjusted sliding scale. Coverage was provided for catastrophic events and services that were deemed appropriate based on national guidelines.

For those so inclined, coverage was available through the nationally licensed insurance carriers that provided benefits based upon the plan chosen. Given the federal requirements for coverage of pre-existing conditions, portability, and guaranteed insurability the reality of adverse selection led to substantial premiums but this option was favored by a significant proportion of the population.

To say it was a tumultuous time for clinicians, hospitals, pharmaceutical and device companies would be an understatement. In retrospect, while still requiring some adjustments, it was the most important payment transformation since the introduction of Medicare in the 1960s.

Cybersecurity

Concerns about cybersecurity initially slowed adoption of technologic advances in sensors, MEMS (microelectromechanical systems) and data analytics. However, the earlier investments made in autonomous vehicle security systems greatly benefited clinical systems. Multiple solutions were implemented including hardening implantable devices, independent authentication applications and private networks. While remaining alert for new threats, cybersecurity concerns have for the most part been mitigated.

Medical education

Although the initial period of medical student education is still termed "medical school" it is highly variable. Schools have differentiated themselves based on their instructional strengths. Some focus heavily on virtual training, others move their students across multiple sites emphasizing immersive experiential learning. Schools further differentiate themselves by their emphasis on preparing students to meet competencies associated with procedural disciplines, to become sensorologists, virtualists, or other disciplines. Since "graduation" is predicated on achievement of nationally established competencies, schools tout their time to competency in addition to the opportunity to qualify for additional certifications while completing their foundational medical school requirements.

Similar to the situation with "medical school," the next step in the educational process continues to be referred to as a "residency." However, the progression from medical school to residency is now based on demonstrated achievement on discipline-based national competency metrics. The tradition of Match Day has ended and now there is a year-round process of linking students with residency training programs. As completion of residency training and progression to unsupervised practice is also based on performance rather than "time on task" there is a continuous inflow and outflow of residents in the discipline-based training programs. The seamless progression from entry into medical school through completion of residency requirements within a single academic medical system has become a dominant model. Unaffiliated residency programs have essentially disappeared.

The rapid advances in scientific understanding, technology and computational sophistication have created a market for online certification programs designed for practicing physicians who desire additional training in an area of interest or need updates to avoid becoming irrelevant. While "continuing medical education" or "continuous professional development" courses had long been offered by professional societies and academic institutions, university-based medical schools are now major providers of such programs in collaboration with faculty from engineering, business, law, public health, and other fields. Indeed, for some medical schools these represent the majority of their educational portfolio.

Accreditation

The residency program accrediting bodies evolved as clinical care delivery was rapidly changing. The tradition of residency programs being governed primarily by practitioners from that discipline ended as technologic and scientific breakthroughs disrupted traditional disciplines. The most contentious problems surrounded the closing of programs and elimination of long-standing disciplines. The adoption of machine learning algorithms that read digitally generated data eliminated the need for radiologists, pathologists, sleep medicine specialists, and others. Several procedurally focused and niche knowledge subspecialties such as oncology were rendered obsolete by technologic advances. Not surprisingly, some disciplines initially tried to maintain relevancy by shifting the focus of their training programs. However, due to the changes in reimbursement policies it was no longer financially feasible for clinical care systems to support such residencies and their faculties oftentimes did not have the expertise necessary to prepare residents for competency standards.

By mid-century, many of the traditional discipline identifiers such as internal medicine, general surgery, and family medicine have in large part been abandoned. At the same time, new programs have emerged including sensorology, prognostication intervention, adult population and chronic disease management, and virtual medicine. Despite the emergence of new disciplines, the overall number of specialties have decreased.

Medical school accreditation also faced numerous challenges. The movement of medical schools to differentiate based on their educational processes and the types of clinicians they produced forced the accrediting bodies to abandon their generalized metrics and develop criteria based on the medical school's core missions. The concept of a generalized undifferentiated medical school graduate who could go in a multitude of directions was abandoned and replaced by focused educational programs that begin in medical school.

While the ideal of a unified accrediting system across the continuum of medical education was not realized, there are now tight linkages between the medical school and graduate medical education accreditation bodies. Perhaps most significantly, the accrediting bodies for residency programs and medical schools now make their accreditation judgments primarily based on the performance of their graduates on nationally agreed upon competency standards rather than process measures.

Certification

Physician certification is an ongoing process. The era of high-stakes examinations such as USMLE and board certification exams has ended. Rather, certification is dependent upon patient outcomes. Initially, issues such as "adverse selection" skewing patient case mix were deemed insurmountable, but sophisticated case mix adjustment algorithms have addressed these concerns. More problematic was the transition from individual physician certification to team-based certification. Recognizing that quality care required a high functioning team comprised of individuals from multiple professions, certification now crosses traditional professional boundaries and is determined by the outcomes of care, not only for physicians but also for other clinical professions, including (but not limited to) nursing, pharmacy, and physical therapy. As can be imagined, this was a major disruptor not only to the certification system but also to traditional professional identities.

Many think that this new system that bases certification on team as well as individual accountability is overly complex. While it is anticipated that further modifications are forthcoming, it is now recognized as an improvement over the previous system that emphasized only individual accomplishments and arbitrary measures oftentimes unrelated to actual patient care. Now physicians are licensed to practice when they have met their discipline's residency graduation competency standards. After their first year in practice they are eligible for certification based on their team's performance, which is assessed on a continuous basis as measured by patient outcomes. If their team merits certification status then the physician's individual contributions are assessed and, if they meet national discipline based standards, certification is granted. Certification is continuous as long as the team's and individual patient outcome metrics are met. A similar process is in place for other clinical professionals involved in patient care.

While in an ideal world this solution would have emanated from discussions among the various certifying bodies, that was not the case. Rather it was the united efforts of patient advocacy groups, payers, and the revocation of recognition of some accrediting agencies by the Federal Department of Education that drove this process. In a departure from the historical view where the profession controlled the criteria for admission or certification, it was the perspective of the patient that ended up driving the revised certification system.

Clinical professionals

While traditional clinical professions, including physicians and nurses, play central roles, physician assistants, community health workers, personal technical diagnosticians, and multiple other disciplines are directly or indirectly providing care. Additionally, online virtual assistants, AI enabled chat boxes, and a wide array of "smart technologies" provide advice, monitoring, and "direct to patient" diagnostic capabilities.

While this exploration of the mid-21st century healthcare landscape focuses on the physicians, it is important to remember that they function in a complex system that is only alluded to. No one can say with certainty what that landscape will entail. Compared to today, some aspects of practice will be unrecognizable, while others will change very little. It is my intention that these scenarios inspire you to envision a different tomorrow.

Now let us now look at a typical day in the life of four mid-century physicians. As you read these scenarios, reflect on the following questions:

Is this a utopian or dystopian picture?
How will patients, society, and physicians balance the tradeoffs between privacy and benefits?
What sorts of scenarios do you imagine for a day in the life of a physician as you think about mid-century practice?

What are the implications for medical educators across the continuum of education? How do we best prepare our students and trainees for the future as you see it?

Dr. Shieh (sensorologist)

The ever louder singing of the wren awakened her from her deep slumber. Realizing that it was her alarm and not an actual bird, Dr. Shieh was again grateful that those noxious alarms from the 20 teens had been replaced by less jarring methods based upon your brain waves rather than an abrupt jolt. Slipping on her biosensor workout clothes, she decided to start the morning with light exercise. The treadmill display suggested that she drink 10 oz of water to hydrate herself while working out. Having had stress injuries as a teenage gymnast, she was pleased with how well her biosensor garments, with their embedded EMG sensors and stress monitors, communicated with the treadmill to eliminate overuse injury and that her shoes adjusted to minimize impact.

As today was a light aerobic workout, Dr. Shieh decided to check in on her patients while on the treadmill. Although her training and certification in Biometric Sensorology allowed her to care for patients with a multitude of chronic diseases, she was pleased with her decision to limit her practice to patients with hypertension. She partnered with the large North Central United Health System to provide monitoring and advice for all their patients with hypertension. Currently she was monitoring over 40,000 individuals.

Flipping on the display screen, Dr. Shieh saw that 359 patients required her attention. Quickly scanning the categories of patients flagged by her analytics, 279 were listed because their therapeutic agent sensors indicated they had not taken their medication as directed. As the prescribed medications were known to be efficacious for each individual's unique physiology and comorbidities, she was confident that if they only adhered to their medications, their pressures would be readily controlled within their individualized ranges. To all but 19, she sent "personalized" automatic notes reminding them of the importance of taking their medications as prescribed. Since none of them had ever met her, the note was from the clinician they were assigned to. As always, she hoped the database was up-to-date and that the right clinician was associated with each of the patients. She would check in later in the day to make certain that both the patient and the clinician had received her message.

The 19 remaining patients required additional attention. Six of them were concerning in that the authentication signal, verifying that the data she was receiving came from that patient, was missing. While major cyber-attacks were largely a thing of the past, she was always vigilant for cybersecurity breaches and was concerned that this might be a signal of such a breach. As North Central United Health System's payments were dependent upon their continuously monitored hypertensive patients achieving individually determined blood pressure control at least 97% of the time, even small data breaches could have major financial consequences. Moreover, she knew that other Biometric Sensorologists would love to get her contract. Looking at the specifics of the six patients, she was somewhat reassured that they came from four different clinical locations. That said, she sent a message to each of the four clinics alerting them to the absent authentication signals, and letting them know that her technical diagnostic team needs to follow up personally with each of the identified patients, pending authorization from their clinician. Given the low likelihood of a significant problem and the importance of early detection if there was a cyber breach, she instructed her robotic assistant to remind her to call any clinic that had not responded within 90 min.

The other 13 patients were a hodgepodge. She recognized five of the names as chronic repeaters. Unfortunately, their situations were complicated by substance abuse. She knew that their clinicians had tried multiple interventions and indeed hypertension control was achieved for months at a time but then problems would arise. Knowing the science behind re-minders, she realized that simply sending them another "personalized" electronic message would be fruitless. Hence, she notified each of their clinicians about the potential recidivism. Hopefully intervention would be successful.

Of the eight remaining patients, there were two patterns that were worrisome. In five individuals there had been a pattern of nocturnal hy-pertension whereby their pressures did not register the expected decre-ment during sleep. As prior studies had shown this to be a marker for potential cardiac events, she contacted their clinicians and recommended that the appropriate proteomic profiles be ordered to assess whether this was an isolated finding or an indication that these five needed inter-vention due to eminent cardiac events. In the remaining three, based on their individual profiles, there was an exaggerated morning spike. While the increase in blood pressure just prior to awakening was a nor-mal phenomenon, the degree of the spike in these three individuals was potentially worrisome and potentially indicative of impending stroke. As she was not privy to all their clinical data, she notified their clinicians about this finding and outlined her concerns referencing the latest data comparing the most recent blood pressure surge tracings to each of the patient's baseline tracings.

Getting off the treadmill following her 40-min workout, she got dressed and prepared for her video meeting with the sales representative for a new model ring sensor. It was intriguing in that the web pitch showed that it could be morphed into many very stylish jewelry patterns and there was a model that would, if caught, be torn off the finger without damage to the finger and yet was sturdy enough that it would function well in usual daily activities. She found this especially appealing as a number of her patients had work or hobbies whereby they were hesitant about wearing a ring sensor and getting it caught and damaging their finger.

Before linking to the video sales pitch, she was pleased to see that the clinicians at the four sites had authorized her to contact their patients re-garding the missing authentication signals. She contacted her technical support company, sent them the data on each of the individuals, and asked them to do remote diagnostics. As always, if remote diagnostics was not successful, she requested that they let her know and she would again contact the clinicians to authorize an in-person meeting with each of the patients by a personal technical diagnostician. As the patients did not know the personal technical diagnosticians, she made it a policy to have the clinic contact the patient before the technical diagnostician's visit and then make sure that the technical diagnostician not only provided iden-tification but also referenced the patient's clinician. She recognized that while the patients knew at some level that a Biometric Sensorologist was involved, it was better if the patient perceived that their personal clinician oversaw their care. After all, despite the advances made in health and disease care, she was still trying to manage human beings with all their foibles.

Following the sensor sales meeting, she decided to order a limited number of the ring models that could be worn by patients engaged in potentially dangerous work or hobby activities. Not only did they appear to be as advertised, she was pleased that the company had run extensive validation tests and that they were compatible with her authentication system. While these models were more expensive than the standard sensors that she had figured into her contract with North Central United Health System, Dr. Shieh knew the importance of staying abreast of new sensor developments to keep her top decile patient satisfaction metrics.

Sitting down for lunch, she reflected on her practice. She recalled her history of medicine class where they were regaled by the professor pointing out that at one time there were recommendations that physicians applied to patients based on their age, concomitant diseases and other incredibly unnuanced parameters. Reflecting on the advances that had been made whereby she could tailor pressure recommendations to the individual based on algorithms that took into account hundreds of variables, she wondered what advances would be made over the next 20 years. Perhaps Biometric Sensorologists would go the way of former specialties like radiology and pathology. But such speculation was only that. In the meantime, she was pleased that the renal, cardiac, and other complications of hypertension were prevented by the care she could supervise. Now it was time for her weekly tennis match.

Dr. Deploratae (interventional prognosticationist)

Population health: Health Enhancement and Disease Prevention Department

Scanning through the full genome reports from the 323 children born yesterday in the State, Dr. Deploratae was pleased to see that only one child had the genotypic signature indicating high risk for developing pancreatic inflammation and diabetes. He quickly sent an electronic message to the Prognostication and Intervention Department of the North Central United Health System where the birth mother and infant were receiving care, asking that an appropriate professional meet with the mother. She needed to be counseled regarding the importance of feeding the newborn the probiotic mixture known to minimize the risk of developing diabetes. As he was doing so, he reflected that while the exuberance surrounding the early years of human genomic analysis was somewhat unfounded, he was always pleased when the newborn screening program identified a predilection for disease development or a genetically determined condition that could be treated to prevent symptomatic disease development. It made him happy that healthcare was finally able to deliver on the promise of health rather than focusing primarily on disease.

As one of the division heads of the state's Health Enhancement and Disease Prevention Department, Dr. Deploratae had the opportunity to interact with health systems around the State. A recurring message was the importance of collaboration and teamwork. While the State had been given responsibility for enhancing the health of and preventing disease in its citizens, the reality was that much of this was done through collaborative efforts with the health systems within the State. This particular infant was just one example. It was only through partnership with the responsible clinicians that his state-level department could ensure that infants identified as at risk through genomic analyses were put on the appropriate diets or given preventive medications and follow-up and that their families were provided with support and counseling on behavioral and environmental aids.

More complex was the coordination and teamwork required to monitor specific segments of the state's population. Advances in understanding of health and disease development led to the understanding that the interactions among an individual's genome, microbiome, behaviors and environment were the preponderant determinants of whether the person would develop disease. At the state level Dr. Deploratae could provide geospatial information regarding environmental risks. But the linkages to genomic, microbiome, and behavioral variables required the involvement of the healthcare systems.

Due to public privacy concerns, the procedure that had been established required each health system to enter the individual's data while the State was responsible for updates of geospatial information and the analytic algorithms. Moreover, when individuals were flagged as being at risk, the responsibility for ongoing monitoring of their microbiota for potentially unhealthy changes, their metabolome, transcriptome, proteome and other biomarkers and the provision of individualized behavior counseling fell to the Prognostication and Intervention Departments of the health systems. While Dr. Deploratae rued the complexity of the System, he recognized the political realities that led to its creation. He also recognized that the State would fail in its mission of enhancing health and preventing disease without the full participation and buy-in of the clinicians and staff in the Prognostication and Intervention Departments of the health systems.

But communication with health system leaders, clinicians and staff was only a small part of his responsibilities and generally the easiest. More challenging, was the ongoing task of trying to make his state's citizens and their elected representatives sophisticated consumers of healthcare information. Unfortunately, the anti-vaccination movement of the early 2000s persisted and had been joined by multiple other groundless health and disease theories, oftentimes celebrity endorsed and promulgated. Sometimes, he wished that social media and the Internet would be banned as they provided the "nut cases" a platform to spread their unfounded ideas. Indeed, he had a staff member devoted to identifying nonscientific claims that were unhealthy and harmful and following their virtual communities online. His department's bloggers and social media communication experts were constantly trying to expose the charlatans but it was an ongoing battle.

As his autonomous vehicle transported him home from another town hall meeting, Dr. Deploratae reflected that despite all the technologic and biomedical advances, the major challenge in his job was communicating with the lay population. How to help them understand that their behaviors and the environment in which they lived, worked and played were just as important as their genome and their microbiota. How to encourage and enable them to make changes in their behaviors that would enhance the likelihood that they remained healthy. All the technologic and scientific advances were for naught if their risky behaviors persisted.

Unfortunately, the public-at–large was not the only group that he had to convince. One of his biggest challenges was educating the elected representatives and his state's bureaucrats. With each general election and administration change, there was a new group that needed to be convinced of the value of investments for disease prevention and the encouragement of healthy lifestyles, that focusing screening and preventive measures was the fiscally prudent action to take. Generally, he was successful but invariably there was a doubter who ascribed to one or more of the nonscientific and baseless theories and wanted the department to endorse and support that movement. Even more difficult was convincing the state's chief budget officer that the health systems were trusted and vitally important links to the citizens. It was the fees earned by these systems that supported the state's mission to enhance health and prevent disease.

Dr. Ne (interventional prognosticationist/administrator)

N of one health: Prognostication and Intervention Department

Leaving her home, Dr. Ne commanded her autonomous vehicle to take her to the state's Health Enhancement and Disease Prevention Department building. Although she was able to work while in transit, she always wondered why the State insisted on in-person meetings rather than using conferencing technology. As the Vice President in charge of the North Central United Health System's Prognostication and Intervention Department, she was required to meet quarterly with Health Enhancement and Disease Prevention Department auditors and review North Central United's prevention metrics as required by their contract with the State. While the activities themselves were reasonable and were consistent with national recommendations, the reporting requirements and audits seemed unduly burdensome.

Having completed review of the performance data she would share with the auditors, she reflected on the past 11 months since her promotion to Vice President. Trained and certified as an Interventional Prognosticationist, her focus was on early disease intervention and prevention. While she knew intellectually that her leadership role magnified the impact she had on people in the System, she missed the fulfillment she felt in her clinical role. It was always gratifying to identify individuals whose biomarkers and underlying genomics, microbiota, behavior and environment predisposed them for problems that could be prevented or indicated that they had early disease that could be reversed. Though she would not work directly with the individuals, her team of Interventionists—the System's coaches—would meet with the patients and work to implement the recommendations she made to prevent disease or to reverse it before it became symptomatic. As she often said, her team members were the ones who made "health" a reality in the North Central United Health System.

Arriving at the Health Enhancement and Disease Prevention Department building, she mentally reviewed how she would approach the meeting. She regretted that shortly after her promotion she had arranged to meet with the auditors and suggested that rather than quarterly in-person meetings, she would submit North Central United's data monthly and if there were questions she would be glad to address them but did not see the need for their face to face meetings. She now realized that governmental bureaucrats were not interested in changing long-standing practices even if it meant enhancing efficiency. In retrospect she wished that she had not been so naïve as their relationship got off to a rocky start. It was but another lesson among the many she had learned since becoming Vice President. She only wished that she would have had the opportunity to be trained and to learn about practical aspects of leadership prior to assuming the role.

Following her meeting with the auditors, during her ride back to her office, she began working on her monthly departmental communication. While she still felt that her old team, with its focus on early disease intervention and prevention, made the greatest difference in the lives of individuals, she had come to appreciate the contributions of the different divisions. She recognized the value of her Interventional Prognosticationist colleagues who provided patients with accurate estimates of survival when they were dealing with cancer, heart disease, renal failure, and essentially all of the common serious diseases. She had learned that the resulting conversations between the Adult Patient Management doctors and their patients, while difficult, were a great improvement over the "average survival" estimates that docs quoted in the past. She had even been told that patients and their families greatly appreciated the realistic estimates as it allowed them to plan accordingly. Directly connecting the day-to-day work of the physicians, Interventionists, and staff of her department with the benefit they provided patients and their families was a goal for every monthly communication as it showed them how meaningful their work was.

Another important goal was educating all members of her department about the diverse skill sets and expertise within the department. She had never really thought about it but assumed that as a clinically active Interventional Prognosticationist she was familiar with the skill sets of the Interventionists. But, since becoming Vice President she had developed a deep appreciation for the full scope of their responsibilities. Skilled in behavior change, counseling, negotiating with patients and their families, they were the effector link with the patients and oftentimes advocates for their patients. They worked with mothers whose infants were found to have a genetic disease or a predilection to disease that required dietary, behavioral or therapeutic interventions. They instructed the patients with chronic diseases, such as diabetes subtype 17A, regarding individual-ized dietary modifications. They worked with governmental agencies to address environmental hazards that were putting patients at risk. They worked with patients to change habits and lifestyles that compromised their health and much more. Seeing this aspect of the System was not only eye-opening but made her proud to be leading such a dedicated group of professionals. Through her monthly communications, she aimed to con-vey not only the importance of their work but also the breadth and depth of skills possessed by the professionals and staff in the department. After all, only through the efforts of every team member were the benefits real-ized by their patients. In her opinion, her department, the North Central United Health System's Prognostication and Intervention Department, was the most important department in the System and she wanted all members of the department to share in this sense of accomplishment and pride in their contributions.

Arriving back at the office, she grabbed a quick lunch while her elec-tronic assistant briefed her on the afternoon's meetings. The first was with a young physician who felt that he was not being recognized appropriately for his intellectual contributions. After hearing the physician's concerns, Dr. Ne assured him that she would investigate them but would not make a decision until all sides had been heard. While this was not the outcome that the young physician desired, Dr. Ne had learned through experience there was often times three sides to a situation—the perspectives of the two parties involved and the truth, somewhere in the middle.

The next meeting involved her division heads and finance. She desired to design and implement a program to reward and motivate the person-nel in the department across the multiple roles. As she had no training in what motivates individuals to perform at a high level she had to learn "on the job." Unfortunately, she realized that her Division Chiefs were similarly ill prepared for this aspect of their jobs. While monetary rewards had always been seen as the only important part of the equation, she had learned that autonomy, a sense of mastery, and appropriate recognition were equally, or even more important than financial considerations for many of the department's members. Moreover, what motivated an individual changed depending upon the stage of their professional career and their life circumstances. While complex, she knew this was critical for the continued success of her department. At the same time, she knew that they would likely not get it "right" the first time and would need to modify the program in upcoming years. In many ways, this was among the most important, and difficult, changes she would make.

When the meeting ended, one of her Division Chiefs stayed to discuss recruiting an additional physician. In this situation Dr. Ne proposed "re-tooling" a current departmental physician to fill the need. As advances in the scientific understanding of health and disease and technology meant that the knowledge and skill sets required of an Interventional Prognosticationist were continually evolving, the desire to recruit additional physicians was a constant. However, Dr. Ne always reminded her Division Chiefs of the realities of financial constraints as well as the importance of providing physicians who had been vital contributors to the department's mission the opportunity to continue doing so by making investments in their ongoing professional development.

Finally having a break from her steady stream of meetings, she was once again struck by the irony of her role. Prior to accepting the Vice President position she had not appreciated that her days and weeks would be filled with meetings, many of which were of the old-fashioned face-to-face variety rather than facilitated by technology. How recruiting, dealing with retention issues, designing and implementing programs to reward and motivate the personnel in the department, and communicating her vision and direction for the department would be so challenging. And most surprising of all, the number and complexity of personnel issues she would have to adjudicate. She sometimes chuckled to herself, remembering the epiphany that struck her early on in this position, that while her training as an Interventional Prognosticationist prepared her to apply the most modern cutting edge science to human health and disease, it did not prepare her for the reality of departmental leadership and management. Success in her current role required centuries-old skills of human-to-human interaction and not cutting-edge scientific expertise. If she had only known, she would have paid attention to learning about organizational psychology, motivation and high functioning teams.

Rousing herself from her reflections, it was time to prepare for her evening meeting with a group of North Central United Health System's supporters and benefactors—it was going to be another late night. Calling up the bios on the attendees, the level of detail that the bio app assembled by analyzing the multitude of digital databases and social media, never ceased to amaze her. As had become her practice, Dr. Ne focused on how best to make a human connection with each—their interests, children and grandchildren, recent trips and vacations—topics that would connect with them at the personal level. Reflecting on the day, she realized that while the focus of the discussions was variable from meeting to meeting, to be successful as Vice President the most important skill sets were not her scientific and clinical expertise but rather her ability to relate at a personal human level.

Dr. Josiah (adult disease management)

As was his custom, Dr. Josiah was the first to arrive at the clinic. Upon entering his office, he biometrically activated all his information systems, giving him full access to the complete records, genomic, biomarker, behavioral, geospatial linked environmental data and sensor recordings for the 15,000 adults for whom he was responsible. He was pleased to see that there were no reported traumatic or other unanticipated events overnight. The weekly virtual meta-team meeting was scheduled to start promptly at 7:30 a.m. so he poured himself a large mug of coffee and joined in the discussion about the upcoming mayoral election prior to the actual start of the meeting. The meta-team was composed of a diverse array of professionals, ranging from medical to urban planning. This was the team that had overall responsibility for the more than 500,000 individuals in the North Central United Health System.

Following the usual sequence, the first update was from the Community Health workers, reporting on an apparent spike in respiratory infections, corroborated by the pharmacist reporting an increase in over-the-counter cold and sinus remedy sales the past 8 days. One of the nurses reminded the team that a few early cases of influenza had been reported in the State so everyone needed to consider influenza as well.

The next agenda item was a review of the 143 orthopedic patients who either had procedures or were scheduled for a procedure in the next month. It was noted that one of the patients scheduled to have a joint prosthesis replacement due to failure to respond to regenerative interventions was not following her prehab as reflected in her sensor readings. It was decided that one of the Community Health workers would make a home visit to encourage her and point out that her surgery would be delayed unless she achieved the requisite level of fitness that optimized a speedy recovery and successful outcome. Two other patients, who had knee procedures, were not progressing in their post-op rehab as predicted. Additionally, their sensors indicated that both continued to have gait abnormalities and limited range of motion of the affected knee. The team decided that a physical therapy assistant would make a home visit to assess each of their living situations and to determine whether there were extenuating circumstances that needed to be addressed. If simply additional encouragement was needed, the team decided to re-enter them in the remote video enabled group PT program. The remainder of the patients were on schedule as reflected in their sensor data and were sent "personalized" computer-generated notes commending their progress and encouraging them to maintain their discipline.

The next agenda item was always Dr. Josiah's favorite. While tremendous strides in analytics and home diagnostics had made the vast majority of illnesses readily diagnosed, puzzling situations on occasion arose where the computer algorithms could not arrive at a diagnosis with the required confidence levels. This was when Dr. Josiah, who held a secondary certificate in Diagnostics, Exceptions and Rare Diseases, was called upon. The most common reason for such a situation was incomplete data or that it was too early in the natural history of the disease to reach a definitive diagnosis. However, about every 2–3 months he had to make a call as to whether they were seeing simply a very uncommon presentation of a relatively common disease or a truly rare disease that was present in such small numbers that the diagnostic algorithms had insufficient input to make a diagnostic call. Though he had never been so fortunate, he always dreamed about what it must have been like to be the first to describe AIDS, toxic shock syndrome and other diseases. When daydreaming, he often thought that the ultimate professional accolade would be to have a disease named Josiah Syndrome just like primary aldosteronism had been called Conn's Syndrome in the old internal medicine books.

While the final agenda item, health enhancement, was arguably the most important aspect of the team meeting from a societal perspective, Dr. Josiah often allowed his mind to wander. He readily acknowledged the importance of environmental, lifestyle, behavioral and community interventions to optimize healthy living. However, he also realized this was not his area of expertise and he had little to contribute other than opinion. Consequently, he left these matters to other team members with the expertise to make recommendations and implement plans.

Upon completion of the virtual meta-team meeting, Dr. Josiah turned his attention to his primary role in the North Central United Health System, managing the 15,000 adults who were his responsibility. As his primary certification was in Adult Population and Disease Management, he oversaw a team of three clinical assistants and received daily information from several Biometric Sensorologists. Together they managed the health and diseases that arose in their panel population.

Given the reported surge in respiratory illness and the fact that influenza season had arrived, he quickly scanned his information systems to see if any of their panel had sensor readings consistent with influenza. Fortunately, it did not appear that influenza had reached their community, at least not in his population panel. That said, he ordered "personalized" computer-generated reminders be sent to all individuals who had not received appropriate influenza vaccinations under the most recent CDC guidelines.

He noted that one of his patients had just been recorded as having a positive home pregnancy test. Her therapeutic compliance database link showed that she had been faithfully taking her preconception vitamins, was on no other medications and was healthy. Seeing that this was her first pregnancy, Dr. Josiah sent her a "personalized" computer-generated note outlining what to expect, links to the online First Baby information and support group and reminding her to make an appointment to follow up with a midwife.

Next, he turned his attention to the automated lung imaging screening reports he received on his at-risk for lung cancer cohort of 63 individuals. The majority had genetic and environmental exposures that increased their risk for lung cancer. However, despite his best efforts and that of the Prognostication and Intervention Department's Interventionists, he still had four people that refused to quit smoking cigarettes and an additional seven that were chronic high-level recreational marijuana users. As he reviewed the machine-read images, he briefly reflected on how disruptive technology had been to the radiology specialty. Radiologists had largely been replaced by algorithms and machine reading technology. But then, other specialties had similarly been affected—pathology, dermatology, even oncology. As he had ready access to all the trials and medical advances in the world through his information systems, he no longer needed the input of many traditional medical subspecialties. The computers not only knew the world's medical literature, the embedded algorithms would tailor it to each patient's case, as needed. He sometimes wondered which discipline would be next but then realized that at critical times in every one of his patient's lifecycles he could provide the human interaction that could not be replicated by a machine, or so he hoped.

While casually scanning through the reports and reflecting on the role of technology in medicine and healthcare, his musing was interrupted by an alert report of a highly suspicious nodule in the left lower lobe. Calling up the patient's full database, he was pleased that there had been no weight loss, fevers or specific proteomic, metabolomic or other bio-marker changes suggestive of metastatic disease. Regardless, follow-up was indicated and the first step was to obtain a blood sample to ana-lyze for circulating tumor DNA or cells. This was when a personalized computer-generated note would not suffice. Dr. Josiah asked his robotic office assistant to contact the patient and arrange a video call to speak directly with him about the finding and his concerns, letting him know that one of Dr. Josiah's clinical assistants would come by that afternoon to obtain a blood sample at a mutually convenient site. Or, if the patient preferred, he could come to the clinic.

While precision medicine had totally changed cancer care, allowing targeted biologic, immunologic and pharmacologic therapeutic agents that were proven efficacious to be administered, it was well recognized that new mutations would arise. Thus, periodic reassessments of the tu-mor's genome were needed. As Dr. Josiah was caring for almost 100 can-cer patients, his weekly routine included review of each patient's status. Depending upon the rapidity with which new tumor mutations were identified by "liquid biopsy," he would decide whether another blood analysis was indicated or, aided by algorithmic analysis of their database, a modification in their therapeutic regimen was needed.

While not "medically necessary," he also made a point to have a video visit with each of the patients on a weekly basis. Since he had all their phys-iologic and biomarker data readily available in his information systems, he knew how they were doing from a physical and even an emotional standpoint. However, to Dr. Josiah, this was part of the "human touch" that none of the metrics captured but was integral to being a physician.

Reviewing his patients with diabetes, he was pleased to see that the fourth-generation bionic pancreas was working well and all had excellent control without any hyper- or hypoglycemic episodes. Moreover, he took great satisfaction that the multitude of complications of diabetes that were all too prevalent back in the early 2000s never developed in the patients with diabetes in his panel.

One of the wiser business decisions the North Central United Health System group had made was to contract with several Biometric Sensorologists. While some were trained as physicians and others came from a biomedical engineering background, all provided excellent ser-vice. Not only did they ensure that data was properly authenticated, they provided the sophisticated sensors and coordinated the deployment of diagnostic technicians if sensor failure or other problems were suspected. Also, as Dr. Josiah had advocated, by dividing the contracts among several service providers, they were not beholden to a single Sensorologist. While it had initially required a bit more technical sophistication to integrate the data streams, it meant that only the North Central United Health System physicians had a complete picture of each person in their population.

Reviewing the reports from the Biometric Sensorologists who monitored his patients with hypertension, pulmonary disease, congestive heart failure, coronary artery disease, his diabetic population, as well as his patients afflicted by some of the less common endocrine diseases, Dr. Josiah noted that all were within their individually determined control ranges. He did have a note from one of the Biometric Sensorologists requesting permission to remotely investigate an absent authentication report. Giving permission, he also authorized a diagnostic technician home visit if that was necessary to resolve the problem.

Having just finished his patient review, he was summoned to the clinic to see a 58-year-old woman who had slipped and injured her forearm. Examination revealed an ecchymosis with attendant tenderness. The report of the scanning image revealed only minor soft tissue injury. Evaluation did not reveal any signs suggesting a neurologic or gait disturbance. He reassured her that the ecchymosis would resolve as would the tenderness. He also gave her an order for gait and balance sensors as part of the workup to see if there was an underlying medical condition that might increase her likelihood of falling.

After a leisurely lunch in the break room, he returned to his office and got caught up on articles that were identified as fitting his clinical interests in diagnostics, exceptions and rare diseases, as well as bicycling and woodworking—his hobbies. All the while, his summary dashboard continued to show green indicating that the members of his panel were doing well.

As the afternoon progressed, sensors indicated that one of his pulmonary patients was beginning to develop problems with bronchospasm and showed a slight decrease in oxygen saturation. Since he had a little time, he called her and asked her to come in so he could examine her. He explained to her if transportation or prior commitments made it difficult for her to come to the clinic he would be glad to send one of his clinical assistants to see her at a convenient time and location. Fortunately, transportation was not an issue and she readily agreed to be seen that afternoon. Intervening early, rather than allowing a condition to progress, was the goal of his practice. Eliminating routine visits, which had been made possible by the sophisticated sensor system their group deployed, meant that he was always available to provide care to patients when their sensors indicated they needed it, sometimes even before the individual recognized the change in their clinical status.

Reflecting on how the practice of "primary care internal medicine" must have been in the past, he was always grateful that his day was not consumed by a constant parade of individuals, most of whom were stable. He could only imagine the frustration those physicians must have felt when they saw a sick patient that needed their care but a waiting room full of individuals meant they could not provide the attention that the sick patient needed.

On his way home that afternoon, Dr. Josiah thought about the day. He had been able to devote the time and attention to the handful of patients who really benefited from his clinical acumen and care. His population panel was stable and those who had chronic disease were within their individually determined control ranges. The meta-team conference had gone well and he had been able to contribute his learned opinion that one of the diagnostic conundrums was simply an unusual presentation of a relatively common problem early in its natural history. Over the next 3 days, he would see whether he was correct. He was worried that tomorrow's report on the "liquid biopsy" of his patient with the lung lesion would show cancer. Fortunately, if that was the finding, the report would include a genomic analysis of the tumor and recommendations for therapeutic agents that would be efficacious, at least initially. It was always good to be able to convey a plan and provide patients some hope when faced with a diagnosis of cancer. Despite so many cancers now being chronic diseases, it remained a devastating diagnosis for so many. All in all, as his autonomous vehicle stopped in his driveway, he reflected that being able to enhance health, and when disease did strike, provide hope and care was a most rewarding profession.

The shaping of "modern" medical education

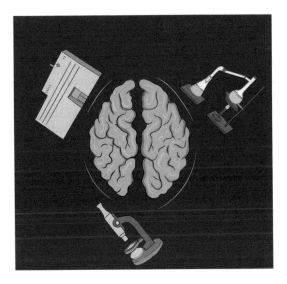

Our perspective on the development of modern medicine and medical education is limited. Regardless of the length of our career, the time we spend in formal medical education and practice is finite. Our viewpoint is further limited by the institution(s) in which we have trained, where we work and the roles we have played.

What we now term modern medicine emerged after the Industrial Revolution. Inventions and discoveries in the 19th century led to groundbreaking advances. During the 20th and early 21st centuries, the proliferation of technologic advances and scientific discoveries continually enhanced our understanding of health and disease and shaped the clinical practice of medicine. Practice patterns have evolved from independent general practitioners to large group practices of employed physicians who encompass a wide range of specialties and subspecialties. The physician's

Implementing Biomedical Innovations into Health, Education, and Practice
https://doi.org/10.1016/B978-0-12-819620-5.00002-3

diagnostic tools have expanded from history taking and patient observation augmented by simple chemical tests and microscopic studies to include sophisticated imaging, biomarker analysis, and a wide range of laboratory examinations. Therapeutic options, initially limited to a few medications and relatively straightforward surgeries, now encompass sophisticated robotic assisted surgeries, organ transplants, implantable devices and an extensive array of medications.

Despite these developments, the basic educational structure and training of future physicians has remained relatively constant since the early 20th century. While the mix of classroom and immersive educational experiences has varied, the 4 years devoted to medical school and the goals of medical education have endured. To develop the foundation for a lifelong professional career, the goals include equipping medical students with the most current scientific understanding upon which clinical practice is based and developing the scientific knowledge necessary to ascertain the "truth" of new discoveries and how these relate to human health and disease. In addition, given medicine's societal mandate, students and trainees are taught to function as clinicians providing the highest quality of care for their future patients.

Often faculty members, students, residents and the lay public assume that medical education and academic medical centers have "always" looked essentially as they do today. This is not the case. The current state of academic medicine and medical education has been shaped over time by the decisions of individual faculty members and by institutional leaders. While all medical schools and residency programs share a common purpose, are accredited by the same agencies and are affected by the same regulations and external events, a given school or program may have responded differently to internal and external realities. Presented here are the generalities. The specifics at each medical school and residency program will vary.

This chapter reviews some of the major factors that have influenced the development of medical schools, residency programs and academic medical centers. While the triumphs of our medical education system are rightfully celebrated, perhaps more instructive are some of the challenges that medical educators encounter. Hence, we will consider the unforeseen educational ramifications of responses to external stimuli and the current challenges presented by financial, legacy, accreditation and certification requirements.

The examples given illustrate some of the many issues affecting medical education. While not intended to be comprehensive, this historical perspective is illuminating as it points out key decision points and shows that the present is not necessarily the future. It also speaks to the importance of thoughtful, forward looking, proactive and sometimes bold rather than expedient, decision making as scientific discoveries, societal change, and the unintended consequences of well-intentioned actions will continue to impact and shape medical education. The lessons of history are instructive. They remind medical educators to be vigilant for changes that might necessitate radical modifications in the preparation of future physicians and to consider the potential adverse educational effects of changes in regulations and practice.

Emergence of modern medical education

As was the norm for many skilled crafts for centuries, learning the medical profession has its roots in the apprenticeship model. A prospective student would apprentice himself to an established physician and learn the craft while earning his keep through menial tasks. As proprietary "medical schools" proliferated in the United States in the 1800s, a common path was 2 years of lectures, the second being a repeat of the first. After graduation, the individual would apprentice with an established physician or begin independent practice.

The staggering Civil War casualties among troops in both the Northern and Southern armies from "medical care," reported to outnumber casualties from combat, led to a national outcry regarding the deplorable state of medicine. There were multiple competing schools of thought, including homeopathy, osteopathy, naturopathy, chiropractic, magnetism, electric therapy, and what was derisively called allopathy by its detractors and scientific medicine by its proponents. While with the benefit of hindsight, we may wonder why the superiority of a scientific basis for medicine was not readily apparent, we must remember that at the time science-based clinical practice was more promise than reality. What we now accept as foundational scientific principles were still being developed and were not uniformly embraced. Surgical anesthesia, through the administration of ether had only been introduced in 1846, the original cell theory proposed in 1838 and modified to essentially our current understanding in 1855, and the germ theory developed and refined leading to the publication of Koch's postulates in 1890. There was a lack of consensus, even among medical school faculty, as to the veracity, let alone clinical implications, of these "new ideas."

It was in this context, beginning in the late 1870s, that a small group of influential leaders of medical education championed the metamorphosis of clinical medicine from an empirical and observational model to a scientifically-based understanding of disease and its treatment. As chronicled by Ludmerer, this resulted in a transition of medical education from lecture-based didactic presentations to an emphasis on foundational basic sciences and laboratory-based experiential learning. Now, over a century later, it is obvious that this set medicine on an unparalleled trajectory of discovery. At the time, however, it led to great acrimony, divided faculties within the schools and led to the eventual closing of many medical schools [1]. Moreover, it was a decades long process that was fully embraced nationally only after the publication of the Flexner report in 1910.

Reflection: Are there parallels now to the situation in the late 1800s? Do societal changes as well as scientific and technologic advances necessitate a reconceptualization of the preparation necessary to be a physician in the mid-21st century?

Development of academic medical centers

Throughout the 20th century, there have been seminal changes that directly affected medical education. A pivotal moment was the embedding of medical schools within universities. Prior to the widespread adoption of a science-based immersive curriculum, freestanding proprietary medical schools were common. Following the Flexner report and the establishment of licensure and accreditation standards, the linkage of medical schools to universities became the norm. To varying degrees, the universities and their medical schools developed associated hospital and clinical facilities that then served a major teaching function.

Research as a priority

Until the 1950s, the focus in academic medical centers was on medical student education. While research was viewed to be important, and the model faculty member was a clinician who pursued research questions arising from the care of patients, it was often times self-funded by the faculty member augmented by modest grants or supported from the institution. During World War II, the federal government's investment in research dramatically increased. This was widely viewed as successful and led to increased investment in the National Institutes of Health and the formation of multiple disease-specific institutes.

The infusion of federal money for research resulted in increased numbers of medical school faculty who devoted significant effort to research. Basic science departments grew dramatically. In many, the focus shifted from preparing medical students to understand, incorporate and practice leading-edge scientific-based medicine to securing grant funding in order to support ever larger laboratories. PhD programs grew and faculty in the basic sciences were doctoral prepared but had little or no clinical experience or expertise. Increasingly, the status of medical schools was measured in terms of NIH funding and faculty performance by the numbers of grants and federal dollars they brought into the school. As grant expectations and the pressure to publish research grew, the demand for increased numbers of doctoral students also increased. The feeling was that not only do doctoral students enhance the intellectual milieu of the laboratory, they also serve as "bright hands" that enhance the productivity of the laboratories and ease the work burden for faculty. An unintended consequence has been a national glut of PhD trained bioscientists for whom there is no readily identifiable traditional academic career pathway [2]. While rewarding and important career opportunities in industry, government, publishing, and other sectors can be built on their doctoral training, all too frequently this is viewed as a "failure" by faculty mentors and adequate time is not provided to prepare for careers beyond the bench.

While the importance of enhancing our scientific knowledge of biologic processes and their contributions to our understanding of health and disease is unquestioned, how best to do so, the metrics used to measure "success" and the appropriate expectations and roles for research focused faculty members in the education of medical students requires thoughtful consideration.

Reflection: Is the distancing of basic science and other research intensive faculty members from the medical student educational mission, as is the norm in many medical schools, appropriate or deleterious to the medical students' education as well as to the faculty?

What should be the expectations, if any, for medical students and residents to be involved in research activities? If this is seen as integral to their education, should the research be directly linked to clinical questions?

Growth of the clinical enterprise

During the first two-thirds of the 20th century, much of the clinical education of medical students and residents occurred in charity hospitals or governmentally supported clinical settings. Often times, the clinical faculty had a separate private practice which supplemented the modest salary received from the medical school in recognition of their teaching contributions. Sometimes the private patients requiring hospitalization would be admitted to the teaching hospital but often on a separate "private ward." Regardless, clinical care was not a major source of revenue for medical schools.

With the 1965 passage of Medicare and then Medicaid, patient care became increasingly profitable and the charitable focus of the clinical practice in many medical schools and their associated teaching hospitals was replaced by an ever-growing focus on "revenue generation." It was soon realized that clinical dollars are fungible and could be used to support both full-time clinical faculty and research faculty. While funding for medical schools throughout the first half of the 20th century relied on students' tuition, philanthropy or, for public schools, from state appropriations, the tripartite mission of medical schools—education, research and scholarship, and clinical care—was increasingly dependent on clinical dollars. As a hospital administrator once told me: "no margin, no mission." This in turn led to a marked expansion in the numbers of clinical faculty, with an emphasis on highly reimbursable procedures and care, and a demand for more residents and fellows who would support the revenue generating clinical activity.

Reflection: The clinical setting is arguably the most important educational venue for medical students and residents. While there has been a synergistic relationship between the teaching and clinical missions for decades, is the teaching mission now subservient to clinical demands?

Unintended consequences shaping medical education

While the overall structure of academic medical centers has remained relatively constant for the last 40 years, multiple factors have continued to impact medical student and resident education. Several emanate from federal policy decisions and the response of institutions to these stimuli. Some, from the real or perceived influence of regulatory bodies, others result from internal institutional decisions. Let us consider a few such examples and the sometimes unintended consequences on the educational process.

Medicare financing

Not only did the passage of Medicare provide impetus for a dramatic expansion of the clinical faculty, it fundamentally altered graduate medical education. During the first part of the 20th century, the need for additional clinical training following graduation from medical school had led to the development of intensive 1-year hospital-based experiences—internships—and generally longer subspecialty oriented but again hospital-based experiences—residencies—that increasingly became the norm by the 1930s. The terms applied to these physicians ("residents," "house officers") were apt as they worked for minimal pay and often lived in quarters provided by the hospital. They were skilled "cheap labor" making up for poor staffing and lack of resources in many city hospitals and charity wards. They were responsible for all aspects of care, ranging from transporting to monitoring and supervising. While the Medicare act led to major improvements in the compensation to interns and residents through the direct and indirect Graduate Medical Education (GME) payments provided by Medicare to hospitals, many hospitals became very dependent on this staffing and revenue source. This resulted in the reluctance to make fundamental changes that would better align learning with professional practice such as moving residency education to ambulatory settings or growing important but low revenue generating residencies. An unresolved tension, exacerbated by direct and indirect GME financing components of Medicare, is the question of whether residents are employees of the hospital who happen to be learning while performing a job or are primarily students who, while learning their profession, also provide a service.

Reimbursement documentation

The symbiotic relationship between learners and clinical faculty maintained the clinical educational mission for many decades. However, another unintended consequence was the Centers for Medicare and Medicaid Services (CMS) (previously known as the Health Care Financing Administration) 1997 decision to disallow attending faculty countersignature of medical student documentation of the history of present illness, past medical history, and physical examination for billing purposes. Prior to its ultimate reversal in 2018, this change, coupled with the implementation of ancillary services for routine medical tasks and the ubiquitous presence of electronic records meant that medical students on inpatient services no longer play central roles and have become increasingly vestigial appendages. Charting, drawing bloods, starting IV lines, tracking down laboratory and radiology results, while menial tasks, allowed the medical students to be integral to the function of the clinical care team. The students often comment that they feel like observers rather than members of the care team. Even though they still do patient write ups, it often seems like "make work" as it is not perceived as contributing to the patient's care.

Similarly, the CMS requirement that faculty must be present for "key portions" of clinical activities, including ambulatory visits and procedures, while meant to ensure the quality of medical care, resulted in residents being given less opportunity to develop expertise, especially in the procedural disciplines. The result has been that learning, skill development and residents' confidence in their abilities has been prolonged and delayed. Clinical activities once seen as primarily medical student opportunities and responsibilities are now the purview of residents; and fellows now do procedures that at one time were primarily performed by residents. This is thought to contribute to the phenomenon of ever-increasing subspecialization, perceived by residents and fellows as necessary for their development of competence, but obviously lengthening their training time.

The tyranny of efficiency

"Reimbursement constraints" coupled with ever-increasing expenses have resulted in growing expectations that clinical faculty will be self-funded and entrepreneurial. While acknowledging the importance of the educational mission, the reality in too many academic medical centers is the administrative leadership's focus on "patient throughput, efficiency, market share, and the financial bottom line." While the most senior clinical trainees, the fellows, along with experienced residents, facilitate clinical care, medical students are often seen as an impediment to "efficient" patient care and flow. Hence, the perspective that deans charged with the medical student mission all too frequently hear from clinical faculty is "my fellows, our residents, your medical students." A recurring lament among dedicated and well-intentioned clinical faculty is that medical student education is least valued among the different facets of education and expertise in education is viewed as less important compared to clinical and research expertise.

Legacy constraints

A less recognized reality is that many academic medical centers have high fixed legacy costs. While the medical science and clinical practice have progressed, the departmental structure and faculties therein are sometimes based on distant past decisions reflecting realities that are no longer relevant. Even though the practice of medicine and science has blurred traditional departmental boundaries, in many medical schools budgets remain departmentally-based and teaching responsibilities for medical students are directly tied to funding. While budgetary specifics differ, the result is that decisions regarding the curriculum are often driven as much by financial considerations as by the needs of the medical students. In many cases, the assignment of teaching responsibilities is made at the departmental or divisional level. When a medical school faculty member's research productivity declines or a clinician's skill set is no longer at the leading edge of practice, they are often given increased teaching responsibilities as a way to fund their salary. When the faculty member embraces the opportunity to educate the next generation, the outcome is positive. However, this is not always the case.

The problem of legacy faculty is not confined to the medical schools. In many colleges and universities, curricular demands generated by premedical requirements have led to sizing of faculty in departments such as chemistry and physics that exceed the numbers needed to teach students majoring in those disciplines. If these premedical requirements were ended, there would be a surplus of faculty in these departments further compromising the financial stability of many institutions.

Accreditation, licensure, and certification requirements

The Balkanization of professional organizations and accrediting and certifying bodies, as depicted in Fig. 2.1, poses additional challenges. Directly or indirectly, each of these groups, and others, play a role in providing and/or

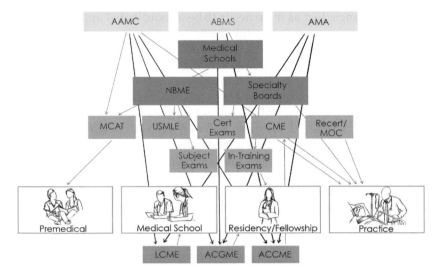

FIG. 2.1   A partial depiction of the complex relationships among accrediting and certify-ing bodies. *Bold lines* signify sponsoring organizations. AAMC=Association of American Medical Colleges; ABMS=American Board of Medical Specialties; ACCME=Accreditation Council for Continuing Medical Education; ACGME=Accreditation Council for Graduate Medical Education; AMA=American Medical Association; CME=continuing medical education; LCME=Liaison Committee on Medical Education; MCAT=Medical College Admission Test; NBME=National Board of Medical Examiners; USMLE=United States Medical Licensing Examination.

overseeing the educational process. Medical schools and graduate medical education programs are accredited at the institutional level, based on insti-tutional and programmatic criteria rather than on the performance of an individual student or resident. Accrediting bodies are charged with main-taining a national level of consistency within a politicized reality. While of-ficially encouraging creativity and innovation, accreditation standards are often perceived by medical educators as constraints to innovation.

In contrast, licensure and certification decisions are based on the indi-vidual's performance. This too presents challenges. Medical educators in-creasingly recognize that their students' and trainees' clinical practices will involve a team of professionals, ubiquitous information technology (IT) support and just-in-time learning/review utilizing frequently updated electronic databases. However, the reality remains that the multiple licen-sure and certifying bodies, while recognizing the rapidly evolving nature of clinical medicine, continue to base their decisions on an individual's performance on high-stakes examinations, generally without the benefit of access to IT resources. This tension between what educators see as nec-essary for future professional success and the way in which their learn-ers continue to be assessed must not only be acknowledged but rectified. This disconnect has the potential to undermine efforts for fundamental curricular change. Just as there needs to be a coordinated approach to the education of physicians across the continuum from premedical to graduate medical education to continuing professional development, so too there needs to be thoughtful consideration given to how better to align the continuum of accreditation, licensure and certification.

A multitude of decisions, whether at a national policy or local implementation level, have had a profound impact on the current state of medical education across all levels. As we consider the path forward, it behooves us to recognize that sometimes well intended measures have unforeseen consequences and that financial demands and perceived or real threats to prestige and power dynamics may impede progress toward meaningful change.

The evolution of academic medical centers will continue. Governmental and oversight agencies will issue new rules and regulations. Our scientific understanding will increase. Physicians' diagnostic and therapeutic options and even where and how care is delivered will change.

Visionary leadership is critical to ensure that academic medical centers fulfill their societal mandate of excellence in medical education. While too frequently we hear "this is the way we have always done it" or "this is just the way it is, there's nothing we can do" such sentiments reflect a limited mindset and timeframe rather than a long-term reality.

Reflection: Given that well-intentioned decisions in hindsight had adverse effects on medical education, how best do we (1) ensure that patient safety is maintained while pushing the limits of students' and residents' clinical capabilities; (2) balance the need for efficiency with the slowdowns that learners impose; (3) ensure that the educational implications of policy decisions are fully vetted prior to implementation; (4) assure the public that national competency standards are being met for physician education at the school and program level while facilitating innovation.

References

[1] Ludmerer KM. Learning to heal: the development of American medical education. New York: Basic Books; 1985. p. 48–57.
[2] Pickett CL, Corb BW, Matthews CR, Sundquist WI, Berg JM. Toward a sustainable biomedical research enterprise: finding consensus and implementing recommendations. PNAS 2015;112:10832–6.

Additional resources

• Ludmerer KM. Learning to heal: the development of American medical education. New York: Basic Books; 1985.
• Ludmerer KM. Time to heal: American medical education from the turn-of-the-century to the era of managed care. Oxford, New York: Oxford University Press; 1999.
• Ludmerer KM. Let me heal: the opportunity to preserve excellence in American medicine. Oxford, New York: Oxford University Press; 2015.

SECTION II

CONSTANTS IN MEDICINE

When interviewees were asked to predict how the practice of clinical medicine would look in midcentury, a recurring theme was the importance of reaffirming fundamental aspects of physicianhood. As thoughtful observers have emphasized in prior generations [1,2], they anticipated that even though there will be a great deal of change in the practice of clinical medicine, there are also many aspects that will remain constant. Indeed, as our understanding of health and disease at its most fundamental levels and modalities to intervene proliferate, it is increasingly important to emphasize the enduring aspects of what it means to be a physician. The profession must remain true to the primacy of caring for the sick and alleviating suffering. This requires the understanding and skills necessary to optimally attend to all the aspects of care beyond simply applying the latest technology.

In this section we will consider subjects that are known for decades or longer to be central to physician practice, but are often much less emphasized in medical education than the foundational principles of biomedical science and the latest applications of technology. The understanding of the importance of all components that comprise a physician's identity and the realization that despite anticipated changes in the practice of mid-century medicine there are fundamental aspects of being a physician that will remain constant is a high priority for medical educators.

Let us look at some of these enduring characteristics.

References

[1] Engel GL. Enduring attributes of medicine relevant for the education of the physician. Ann Int Med 1973;78:587–93.

[2] Peabody FW. The care of the patient. JAMA 1927;88(12):877–82.

3

Healer and professional roles

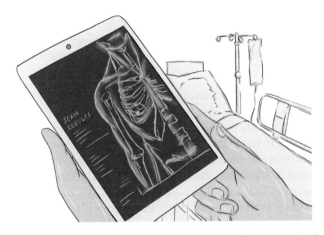

The role of the physician rests on a three-part base, as is depicted in Fig. 3.1. The first is a deep understanding of the biomedical sciences and relevant technology as applied to health, disease prevention, and amelioration or cure of disease. The second is the contract with society as encoded in professionalism. The third is the interaction with individuals in the role of healer, bringing together medical professionalism and the understanding of scientific knowledge and innovation. Importantly, the role of healer includes but is not limited to curing as "restoring to soundness" is a desired goal even when cure is not possible.

The distinction between the physician's two roles is blurred. As expressed by Cruess and Cruess "Physicians fill the role of healer and professional simultaneously, roles that have different origins and traditions. The healer, which is what individual citizens and society require, comes to Western culture from the Hippocratic tradition. The role is reasonably well understood and has had an important place in the medical curriculum for a long time. Professionalism, on the other hand, arose in the guilds and universities of the Middle Ages but had little impact on society until modern scientific medicine developed. As the delivery of healthcare became increasingly complex, Western society chose to use the concept of the profession as

Implementing Biomedical Innovations into Health, Education, and Practice
https://doi.org/10.1016/B978-0-12-819620-5.00001-1

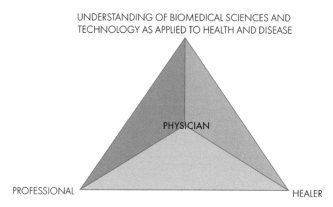

UNDERSTANDING OF BIOMEDICAL SCIENCES AND
TECHNOLOGY AS APPLIED TO HEALTH AND DISEASE

PHYSICIAN

PROFESSIONAL

HEALER

FIG. 3.1   Understanding of biomedical sciences and technology as applied to health and disease. The identity of physicians rests on three foundational components.

a means of organizing the delivery of health services... As industrialized societies became wealthier, the professions were granted status, prestige and substantial rewards on the assumption that professionals would be altruistic and moral in their day-to-day activities. This formed the basis of the social contract between medicine and society" [1]. Thus, the physician has a societal contract as a professional and an individual focus as a healer.

It is noteworthy that even after almost 2500 years, most of the concepts that are embodied in the original Hippocratic oath [2], remain relevant and are found in modern pledges. For example, the World Medical Association Declaration of Geneva the Physicians Pledge [3] includes:

• respecting the autonomy and dignity of patients;
• maintaining respect for human life;
• not permitting considerations of age, disease or disability, creed, ethnic origin, gender, nationality, political affiliation, race, sexual orientation, social standing, or any other factors to intervene between my duty as a physician and my patients;
• practicing with conscience and dignity and in accordance with good medical practice.

But, while these principles of physician conduct are immutable, the clinical context in which they are practiced continues to change. Although this book is predicated on the impact of biomedical and technological advances, it is important to remember that human-to-human interaction is at the core of clinical medicine.

"People need time to heal, in my opinion, and I think we're headed towards the very scientific piece where we're really getting on top of it, I mean we're going to be able to do testing and figure out what medications people are going to respond to, what they're going to react to. We're going to get that down pretty good. My vision though is that we're going to spend most of our energy in developing systems, then we're going to struggle with the thoughtful and caring person being integrated into them and that's going to be, in my opinion, a huge challenge."

Doug Paauw

Patient or person centeredness

Excellent medical care depends upon the physicians, nurses and other professionals in our clinical systems, but they are only part of the equation. At one time there was the presumption that physicians knew what was best for their patients and that the patients would, without question, follow their physician's recommendations. If it was ever true, it certainly is no longer the case. At its most fundamental level all scientific advances, diagnostic and therapeutic options are for naught if the individual and/or surrogates choose to not follow recommendations. As the patient is the central figure in determining whether the ministrations of the clinical care team are efficacious, physicians must not only advise their patients as to diagnostic and therapeutic options, but actively engage their patients in the decision-making process. Understanding the considerations that enter into a patient's decision-making process and eliciting their values and preferences is the essence of "patient centeredness" and optimizing interventions. Determining what the individual patient sees as a "good outcome," which may be incongruent with what the physician thinks is optimal, is central to the whole concept of patient centered care. As defined by a report from the Institute of Medicine (since renamed National Academy of Medicine), patient centeredness is "care that is respectful of and responsive to individual patient preferences, needs, and values, and ensuring that patient values guide all clinical decisions" [4].

While I will expand on the value of the patient-physician interaction and what contributes to its therapeutic potential, it is important to remember that much of what is written and discussed on this topic is from the medical professional's perspective. If we are to be truly patient or person centered, interactions with the clinical enterprise must be viewed from the patient's perspective. What is important will change based on generational viewpoints as well as severity of disease. For self-limited conditions and prevention interactions, convenience and efficient service that minimally disrupts the individuals schedule may be the highest priorities. For life-threatening or life-altering conditions with significant attendant morbidity, patients may place greater value on continuity, expertise, and working with clinicians with whom they can relate. Even for such individuals, convenience and minimal disruption to their daily schedules may be priorities. Given the advances in technology, it is highly probable that patients will expect the same level of service as is the norm in many other industries. Implementing the significant changes in our clinical delivery systems to provide "care that is respectful of and responsive to individual patient preferences, needs, and values, and ensuring that patient values guide all clinical decisions" will require far-reaching structural and attitudinal changes. Consumerism will not exempt medicine.

"...as medical training and care is moved from the paternalistic model, I think the next phase that we need to be paying attention to is the role of physicians in helping to clarify what patients are thinking about, what's in their mind and what they're looking for. It's been very handy and convenient to say, 'well I should listen to what patients want,' but so often patients don't know exactly what they want and are even more confused by what they need … we've often taught our learners to try and read between the lines, listen to what isn't being said. I think we're going to have to take that to another notch because with so much out there that patients can be educated about, information that they can access, they're often coming into our offices, into our care settings with certain assumptions. All too often I see the medical system responding to those assumptions as there's something about challenging those assumptions that is antithetical to patient centered care. But I think as we try to push the art of medicine and the art of care to the next dimension, I think we're going to have to get much better at trying to dissect out what are our patients' understanding of things. They have opinions and triggers that are bringing them to seek care in the office or dig into the computer or to ask a colleague, but they're all based on certain assumptions and somehow, we've got to train ourselves, and our learners to work with our patients and get down to foundational assumptions as to what's going on."

Joseph Kolars

While this is the ideal, the current reality is that we, as physicians, and the systems in which we practice frequently fail our patients. Eliciting individual patient's preferences, needs, and values requires time, communication skills and commitment on the part of the clinician. Moreover, patients are often reluctant to confront their physician regarding inattention to their preferences, opting instead to seek care elsewhere or ignore recommendations. An indication of the magnitude of this problem is revealed by a recent study that used artificial intelligence (AI) to analyze over 500,000 de-identified posts on public online forums for patients living with osteoporosis, psoriasis, psoriatic arthritis, Crohn's disease, diabetes mellitus, epilepsy, rheumatoid arthritis, breast cancer, Parkinson's, and Alzheimer's. Analyses showed top concerns regarding physicians were: lack of communication, including a sense that the physician was not listening; feeling uncomfortable about asking certain questions or voicing their concerns to their physician; distrust of the doctor's motivations for choosing a specific treatment or medication; not knowing what to expect from the treatments and medications; and not understanding what test results mean [5]. Thus, while the person or patient centered approach to medicine has been embraced, it is apparent that many patients with chronic diseases, at least those who are active on these social media forums, do not perceive that key components of patient centeredness are included in practice. Does this mean these individuals are receiving suboptimal care? What are the implications for their willingness to follow their physicians' recommendations? Will these concerns result in poorer health outcomes? Readily apparent is the importance of communication and the development of trust to the realization of the goals of patient or person centered care. We will consider further the importance of communication across multiple modalities.

"How do we create the kind of environment for shared decision-making or shared accountability or even establishing what people's goals are? .... I'm talking about the capacity to really think about what are all of the challenges that you're confronting at different points in the life course, how are they manifesting for you and how do we help you to early anticipate, prevent them, or early anticipate and identify and know how to manage them so they don't become big problems?"

Mary Naylor

Expanding the focus from the patient-physician interaction to the totality of patient or person centered care, students and residents must understand their patients' reality when dealing with multiple chronic conditions.

As stated in a recent report from the International Minimally Disruptive Medicine Workgroup, "Living as a patient presents numerous demands: seeking help by choosing and enrolling in health insurance that they can afford; overcoming frustrating administrative barriers; adjusting to arbitrary schedules and communication gaps; traveling to and participating in appointments; making sense of instructions, test results, and bills; obtaining and renewing prescription drugs; paying bills; and more. ... guideline-concordant care overwhelms patients with multimorbidity and their caregivers; they, in turn, opt out of some or all of their prescribed treatments. This nonadherence affects patient outcomes, wastes resources, and complicates patient-clinician relationships, adding work to their encounters" [6].

These realities—termed the burden of treatment—are too often minimized or not even considered by the patient's physicians.

Further complicating the situation, patients are frequently put in the difficult position of having to decide among what may be conflicting recommendations proffered by different specialists based on the physician's focus on a disease-specific guideline or care pathway rather than the totality of the patient's problems [7].

If physicians and clinical care systems are to be patient centered in the fullest sense of the word, should we as medical educators not only emphasize the importance of actively engaging patients in decision-making, but also help our students and residents understand the ramifications of living with a chronic disease and what their patients experience? And, if we are truly to embrace patient centeredness as something other than a theoretical construct a realignment of our "efficient" clinical care systems may be needed. Patient centeredness is built upon a trusted relationship that requires continuity and time. Both are increasingly rare in "modern medicine" [8].

The continued march of consumerism, the growing array of diagnostic and treatment options, the Balkanization and complexity of our clinical care system and the ready availability of information requires that clinicians consciously seek to understand the priorities of the patient, minimize the hassle factor and burden of treatment, and coordinate diagnostic and treatment plans appropriately. As options continue to proliferate and the ability to intervene much earlier in the disease process to restore health or minimize morbidity increasingly becomes a reality for a broad spectrum of diseases, the importance of aligning the individual patient's and clinician's priorities to enhance the likelihood of achieving a mutually desired outcome increases.

But this is a two way street. Often times the perspective of patient centered care implies that the clinician must include the patient as a partner in care. With the advent of the Internet and social media, global networks of individuals with specific diagnoses have formed. For individuals with rare diseases, this means that they can now inform their clinician of the experiences and observations of multiple similar patients.

"What we've seen a lot in terms of autism is the power of the patient. This idea of networking can really be turned on its head to advantage researchers and clinicians. What we call locus heterogeneity is probably the rule for a lot of diseases, which means that you've got a very specific diagnosis, whether it's autism or breast cancer or you name it, but there's a whole bunch of variants.

I often use this example, we estimate now for autism that there's about 400 to 500 genes that when mutated would actually get to an autistic state in a child. If you are a clinician and you're seeing patients for 30 years, what does that mean? That means it's likely that you've never seen that same genetic cause of autism more than twice in your full 30 years. That means clinically you just don't have the (knowledge) base of seeing the 10th and the 12th patient with a specific genetic lesion.

But what happens now, and what we've been able to leverage, is this idea of the patient network. ... you can identify a network with those individuals that have the same common supposed genetic etiology ... if you have clinicians connected to those patients or patients connected to those clinicians in a network, they now can compare notes across continents on these indexed cases. And what has happened over and over again is we've seen these sub phenotypes have arisen and it's very specific and it correlates with that specific lesion."

Evan Eichler

Ensuring that our clinical care systems and the professionals in them practice in a patient centered way is arguably more important than ever. It is not only central to the physician's role as a healer, but physicians will also benefit from the observations and information that patients individually and through networks can bring to the patient physician relationship. In essence, patient centeredness is respecting and accepting the patient as a partner, indeed, the most important partner in their care.

Potential learning enhancements

- Early in their medical education consider linking a student(s) with a patient with chronic disease and/or disability and their family/caregivers.
  - o Have the student explore the patient's expectations and goals for treatment and through periodic visits over the course of one or more years how goals evolve.
  - o Have the student explore with the patient and their caregivers the frustrations that they encounter when interacting with the clinical care system and their physician(s).
- Ask students to identify a disease condition website and assess the issues and questions discussed on that forum that may not be raised or addressed in the clinical setting.

Additional resources

- Agency for Healthcare Research and Quality. The SHARE approach, 5 essential steps of shared decision making. https://www.ahrq.gov/professionals/education/curriculum-tools/shareddecisionmaking/index.html.

Trustworthiness

When I asked Dr. Sanjay Gupta to reflect on what were universal characteristics of "healers" across cultures around the globe that he considered essential to maintain, he replied "trustworthiness" and acting as an "honest broker."

"There are extremes when it comes to the perceptions of medicine and healthcare, and perceptions of clinicians and practitioners and everybody; nurses, doctors, everybody. On the one hand I think they are highly respected, and there is a remarkable amount of trust. People will obviously come at the most vulnerable times of their lives. They will hand off their loved ones and ask that the practitioner take that person literally to the point of death, hopefully to return them to a better shape than before. It is remarkable, I don't think there's anything else like it and I think there can be nothing going forward, nothing that would potentially jeopardize that aspect of it."

Sanjay Gupta

While advances in science and technology will likely result in many changes in diagnosis and treatment, and hopefully even prevention of a disease, suffering and disability will remain a reality and individuals will continue to seek the help of clinicians. It is precisely because of that vulnerability that trustworthiness is not only important but essential. As expressed by Barnard, "When we turn to health professionals for help in meeting our needs, especially in the setting of an illness that compromises cognitive and deliberative capacities and the emotional equilibrium necessary for the thoughtful weighing of options, we are in a state of involuntary dependence. The relationship at its inception is not one between free, equal, and rational contractors.... Involuntarily needy and dependent— that is, vulnerable—we *trust* that health professionals will:

- *not exploit* our vulnerability for their own self-interested ends,
- *not increase* our vulnerability through paternalistic or degrading forms of help that perpetuate dependency or undermine self-esteem,
- *reduce* our vulnerability by
  o alleviating sources of vulnerability related to disease,
  o aligning with others to alleviate sources of health related vulnerability aggravated by social, political, or economic arrangements,
  o enhancing our capacity for self-determination by preventing, eliminating, or reducing limitations related to disease or its treatment" [9].

There are three relevant facts:

- "Being ill is a radically different state of affairs from being well....
  The underlying thought in illness for most persons, even with trivial and certainly with important symptoms, is: *Is this the beginning of the*

end of my existence? The fragility of our human existence comes before us bluntly when we experience illness. We have *therefore, in the fact of illness, a wounded state of humanity"* [10]. The ill person is inherently vulnerable.

- Implicit in physicianhood is the willingness to help and the intent to use knowledge and skills for the patient's good, not in self-interest. "The physician-patient healing relationship is of its nature an unequal relationship built on vulnerability and on a promise" [10]. Illness reduces the capacity of the individual to act as an independent agent and requires them to trust themselves to the care of another. Thus, they are not only vulnerable due to illness but also to the need to defer to the clinicians, often strangers, involved in their care.
- The clinical decisions and actions taken by the physician and clinical team on behalf of the patient will be scientifically sound and based upon best available evidence. Decisions must be made that are appropriate for the individual. This requires understanding that there is a wide diversity of patient and family expectations, ranging from involvement in clinical decision-making [11] to extent of care near the end of life to expectations for "healing interventions." While potentially considered a quaint notion, trustworthiness is a foundational tenant upon which the clinician's legitimacy is based.

Having established the immutable basis for trustworthiness as essential to the physician's role as healer, what are factors that threaten it in the eyes of patients?

An insidious threat to trustworthiness frequently not recognized by the physician is the perspective, preference, or bias we all have based on experience. Whether conscious or not, this is a reality in clinical care. Based upon the training of the clinician, they will tend to view a specific approach to care as preferred. At its most basic, a surgically trained physician will tend to see surgical approaches as preferred to nonsurgical, after all that is where their expertise lies. Recognizing one's proclivities and how they might not be in the best interests of a patient is inherently difficult. However, it is part of ensuring that "the clinical decisions and actions taken by the physician and clinical team on behalf of the patient will be scientifically sound and based upon best available evidence" [11]. The education and training of clinicians must include reflection on how their personal experiences and expertise will affect their clinical judgment and decision-making, and potentially compromise their ability to act in the best interests of their patients.

Recognizing that trustworthiness is fundamental to the physician's role as healer, a patient's trust in the recommendations of their physician is influenced not only by their personal interactions but also the well-publicized lack of professionalism exhibited by some physicians.

"I also think that despite this high regard for the field of medicine, there is a distrust of modern medicine, more than I think I've ever seen before. I think there has been an erosion of trust over the last 10 years that has been fairly profound and I don't know all the reasons for it. I guess it is multi-factorial as people like to call it, but I think whenever we hear about things like price gouging and things like that, besides just being a bad story, it really, really erodes that trust. It's something we cannot as a medical establishment tolerate because the intangible impact is, I think, enormous. It's hard to measure but I think it's enormous. So that trust part has to be restored and I think that has to be a big part of a future doctor's education."

Sanjay Gupta

To realize the potential of the healing relationship, patients must have confidence that their interests are a priority over the personal interests of their physicians, the clinical corporation or the "medical industrial" complex. The ever-growing onslaught of "direct to consumer" medical marketing adds further confusion [12]. Medications, devices, diagnostic tests, all are pitched on a regular basis through mass media. The potential benefits are emphasized, the side effects are downplayed, and with a standard "consult your physician," the listener is left with the impression that somehow it is a panacea for whatever condition it is targeted. When the patient raises the possibility of obtaining a prescription, having the diagnostic test or procedure, the physician is often left in the difficult position of either acquiescing to the request or attempting to explain to the patient why it is not in their best interest to take or have done what is being marketed. Unfortunately, the power of marketing and social media is such that the patient may react negatively to a denial of their request by the physician.

Safeguarding the good of the individual, the healing role, and upholding their societal contract, their professional role, will be increasingly complex and challenging for the physician. The continued development of mega systems that emphasize the business of medicine, societal fears regarding privacy, governmental control and "rationing," and the specter of the clinician's personal gain overriding the patient's interests—conflicts of interest [13]—all combine to further erode trustworthiness. While attention has been focused on clinicians' conflicts induced by relationships with external business interests, the potential of conflict is inherent in all transactional financial payment models. While the public concerns about withholding care have been discussed and well documented when capitated risk-based reimbursement systems are in place, all financial models that prominently feature an activity-based incentive result in potential financial conflict of interest. As part of their professional role, it must be explicit that the physician's primary responsibility is not to the system, the corporation, or the state but rather to the patient. Not to do so will erode trust and further compromise the therapeutic value of the patient physician relationship.

"I don't know if we've done as good a job as we need to in educating young physicians about the moral responsibilities of the practice of Medicine, that the patient comes first. There are still a lot of tremendous excesses. I have certain rules about my engagement with industry and most of the time I won't even get involved with a clinical trial unless we own the data at our academic research organization. But every now and then I'll agree to be on the steering committee for a trial that's being coordinated by what I call a rent-a-doc steering committee and when I see what's going on it makes me even more concerned. Medicine has become an enormously big business. We (physicians) have to be the honest brokers."

Steven Nissen

As expressed by Pellegrino, "Deprofessionalization is a constant danger in the industrialized and commercialized models of care so many favor today. There is constant danger to the whole idea of a profession in the individual's ordinary urges to power, prestige, riches, or pleasure. Today's prevailing moral skepticism, relativism, and self-interest can easily disturb the integrity of even the most sensitive professional moral compass" [14].

In response to changes in medicine and healthcare delivery systems, there has been increased attention paid to the importance of professionalism [15]. However, the scope of encompassed behaviors and attributes are broad and reflect the culture and societal exigencies within which they are developed. Given the importance of professionalism, we will consider the ethical dimensions later in this chapter.

Potential learning enhancements

- View examples of "direct-to-consumer" advertising and discuss implicit and explicit messages conveyed in the marketing. In small groups have learners develop responses to patients desiring the treatment but for whom it is not indicated.
- Videotape patients willing to share their stories of positive and negative interactions with physicians and their perspectives as to how they judge whether to trust a physician's recommendations. Ask the learners to imagine themselves in the patients' situations to understand their views.
- Many patient stories are available on the web. View and discuss the issues raised by the patient's story and implications for the patient-physician relationship.

o Patient story resources:
   http://www.ihi.org/education/IHIOpenSchool/resources/Pages/
   ImprovementStories/default.aspx
o https://www.mayoclinic.org/patient-stories
o https://www.patientsafetyinstitute.ca/en/toolsResources/
   Member-Videos-and-Stories/Pages/default.aspx

Honest broker

While we have focused on the trustworthiness of the clinicians caring for patients, embedded in this concept is the physician's role as "honest broker." To function as an honest broker requires the physician to reliably judge the veracity of data and how it is interpreted. Given the rapid advances in scientific knowledge, the reporting of sometimes conflicting results, and the fact that even highly motivated clinicians do not have the requisite level of expertise to critically judge the vast amount of available scientific information, it is vital to be able to discern what is credible and what is not. Increasingly clinicians, like laypeople, receive information from multiple sources ranging from blogs to peer-reviewed scientific publications. Online sources provide electronic access to peer-reviewed information and perspective from true experts sharing their insight and knowledge, to charlatans touting unsubstantiated false information. The fact that falsified studies evade detection and are published in peer-reviewed journals emphasizes the importance of critical assessment regarding "whom to believe" [16].

Even more concerning is the growing distrust of science among the population. Unfortunately, the media-induced hype over "discoveries" that prematurely tout breakthroughs for the treatment of cancers, cognitive decline and dementia, and other conditions leads to cynicism and heightens this distrust. The Academy and clinical care organizations have contributed to this. It is magnified by the pervasive presence of unfounded claims that circulate on the Internet and even when shown to be false, achieve a following by those who either do not know or choose to ignore the overwhelming evidence to the contrary, due to their distrust of science and of clinicians.

While the knowledge base of physicians has historically been acknowledged as superior to the laity, the public at large and even patients increasingly view clinicians as simply providing another perspective. Hence, the importance of the "honest broker" role will only grow. Given the volume of information and misinformation readily available to patients, the physician must have the background and gravitas to be recognized as a trusted source, an honest broker of best evidence. If the clinician has established a level of trust with their patients and is thoughtful in communicating the basis for their recommendations, the potential of countering "fake science" will be enhanced. Studies of lay people's evaluations of individuals purporting to provide scientific information have found three dimensions—expertise, integrity, and what is often turned benevolence or an impression of their ethical and responsible characteristics—to be important in judging the credibility of a source [17].

As observed by Sanjay Gupta, trustworthiness, and acting as an honest broker of information, is a universal characteristic of 'healers' across cultures and geography. If anything, its importance will grow for clinicians practicing mid-century. Active attention to maintaining and enhancing the trustworthiness of physicians is an imperative. It is the foundation upon which patients will base their willingness to accept their diagnostic and therapeutic recommendations.

Potential learning enhancements

- Watch in small groups and discuss Kathryn Schulz's 2011 Ted talk "On Being Wrong" https://www.ted.com/talks/kathryn_schulz_on_being_wrong?language=en
- Identify and review popular blogs and websites that promulgate scientifically unsubstantiated perspectives on health and disease. Analyze the basis for their popularity.

Additional resources

- National Academies of Sciences, Engineering, and Medicine. Trust and confidence at the interfaces of the life sciences and society: does the public trust science? A workshop summary. Washington, DC: The National Academies Press; 2015. https://doi.org/10.17226/21798.

Ethical issues

Ethical considerations have been part of the practice of clinical medicine for centuries. The first meeting of the American Medical Association (AMA) in 1847 adopted a *Code of Medical Ethics*. In the intervening years it has been modified and now includes sections on: patient physician relationships; consent, communication, decision-making; privacy, confidentiality, and medical records; genetics and reproductive medicine; caring for patients at the end-of-life; organ procurement and transplantation; research and innovation; physicians and the health of the community; professional self-regulation; interprofessional relationships; and financing and delivery of healthcare [18]. It serves as a basis for the social contract between physicians and society. But guidance on ethical issues involving physicians is not the sole province of the AMA. Virtually every professional organization has issued codes of ethics and guidance regarding ethical conundrums encountered by physicians within their specialty.

"The ethical dimensions of roles I think cannot be ever over estimated. This notion of being an honest broker challenges every one of us as we're thinking about interacting with our students, with our own clinicians, it's a big deal. … you have to think about how we create clinicians with an ethical framework that they can use in all of their encounters. It's tackling really very important big issues.

It means that every case students are talking about should have some questions about whether it challenges any ethical framework, and we don't do that."

Mary Naylor

From an educational standpoint, the incorporation of ethics into medical student, resident and PhD curricula has provided a forum to discuss these frequently encountered, potentially uncomfortable, issues. Similarly, the disclosure of conflicts of interest as a routine part of scholarly lectures and publications has heightened awareness about the potential of financial and similar considerations influencing perspective and biasing conclusions. However, as therapeutic and diagnostic options have increased, as clinical medicine has evolved into the medical industrial complex now accounting for over 18% of gross domestic product, and as societal expectations for physicians have changed, the types of ethical challenges have proliferated and it is likely that this trend will accelerate. Not only will existing ethical challenges remain but new ones will emerge. Let us consider some ethical conundrums that will likely arise in mid-century medicine.

Technology: While the benefits from robotic assistive devices designed to maintain independence and allow aging in place are readily apparent, the potential ethical considerations have received less attention. For example, the benefits of reminding individuals to take their medicine, exercise, and eat nutritious meals are undeniable but the autonomy, and the privacy, of the individual must be considered. When caregivers or visitors are present, will the robot monitor or even police their activity? [19] An argument is made that limiting current autonomy is appropriate in the hope that future autonomy will be augmented. But, is there a threshold or balance point where the hope for future autonomy is so unlikely that limitations on current autonomy become unethical? The importance of allowing individuals to make their own choices and not be subject to different standards due to their age needs to be recognized. The European Union's project ACCOMPANY (Acceptable robotiCs COMPanions for AgiNg Years) has developed a framework for an ethical approach to these issues [20].

The ability of a social robot to decrease social isolation and loneliness through engagement of an individual in stimulating activities has been demonstrated. However, the question remains the degree to which it can simulate and substitute for human interaction. While robots may be able to judge and even emulate human emotion, the question of the appropriate deployment of these devices will remain. Will this technologic capability lead to discrimination based on age resulting in even greater isolation of the elderly? As robotic assistive and social devices proliferate, potentially decreasing the need for clinician intervention and interaction, the loss of the social value of a visit with a clinician also cannot be ignored. Is there a responsibility for physicians to ensure that human contact is maintained and social isolation minimized? [21].

Facial recognition technology is increasingly utilized for security clearance identification purposes. Coupled with machine learning based algorithms, this has clinical applications such as diagnosing genetic disorders with morphologic characteristics. A recently published pilot study using photographs of 32 patients with Turner Syndrome and 96 age matched controls produced an algorithm with 69% sensitivity and 88% specificity. In comparison, 21 attending physicians had a significantly lower sensitivity averaging 57% and an equivalent specificity of 81% [22]. The potential for facial recognition technology to aid in the diagnosis of rare genetic disorders with morphologic characteristics is obvious. However it is also proposed as a means to monitor and provide information related to areas as disparate as behavior, aging, and pain levels. As with many new technologies, not only is there the potential for great benefit but also the reality of major ethical issues. Would it ever be possible to truly de-identify facial biometric data? In order to improve the accuracy of such algorithms ever larger data sets are needed. How will informed consent be obtained when the future purposes for which a person's facial template is utilized may not even be known? What should be the role of physicians in the ethical discussions regarding the purposes, ranging from research to clinical applications, of such biometric-based technology and their advice to patients? [23].

As genomic and biologic data collected in the course of clinical encounters or through commercial entities and data from biometric studies, robotic devices and sensors is aggregated to develop ever more sophisticated artificial intelligence and machine learning algorithms, the significant issues around privacy must be addressed. Do individuals "own" their data? Should individuals be allowed to "opt out" or must they "opt in" for their daily functioning routines, physiologic information, and compliance metrics to be entered as part of a massive database? The value of data is incontrovertible and as the number of participating individuals increases, so will its commercial value. Should there be a financial or other form of acknowledgment (or compensation) for contributions made by each individual? Will physicians and clinical care organizations profit financially by encouraging their patients to contribute their data? Multiple scenarios are readily imagined. Many are fraught with ethical issues not the least of which is major conflict of interest. Objections are already being voiced to amassing genetic and other biologic data for research purposes and potential enhancement of clinical care. It is likely that issues around commercialization of data and balancing societal good with personal privacy concerns and "ownership of one's personal data" will only increase.

Technology will allow individuals with rare diseases to access specialty physicians with expertise specific to their disease regardless of geographic separation. Patients will have the option of receiving care in their homes or wherever it is convenient rather than in person. Physicians and other experts will be available online for a wide range of services ranging from general advice to patient specific tele-consults. Telehealth will provide additional challenges. While there certainly are readily apparent upsides, the ethical issues raised must also be considered. Regardless of format, the physician's fundamental professional responsibilities do not change. Putting the patient's interests first—minimizing conflicts of interest and bias; ensuring that the information provided is accurate and objective— competence; transparency as to the limitations imposed by the technology; and very importantly privacy and confidentiality [24].

Looking a bit further into the future, additional potential ethical conundrums are envisioned. As our scientific understanding of health and disease grows, what should be the responsibilities of the physician or the clinical care system vis-à-vis precursors of disease? We know that behavior, and environment, including social determinants of health, account for many of the risk factors in the development of disease. What will be societal expectations for the role of physicians making recommendations regarding modifying behaviors that have the potential to contribute to symptomatic disease? Just as it is now the responsibility of physicians and clinical systems to follow up with individuals who have missed follow-up appointments or scheduled screening examinations, in the future will it be the expectation that they follow up on dietary recommendations to prevent or reverse disease?

Advances in scientific technology often times precede societal discussion and consensus regarding the advisability of deploying a given technique. Unfortunately, conversations regarding ethical considerations are often reactive rather than proactive. For example, recombinant DNA technology and transgenic animals, now widely accepted scientific tools, initially resulted in considerable public concern which led to thoughtful discussions and appropriateness guidelines. Now technologies such as gene editing with CRISPR, not only have great utility as scientific tools, they will potentially enable devastating diseases resulting from single or a relatively few genetic alterations to be cured. Already applied to humans,

CRISPR and similar technologies may be used to ensure desired traits in "designer babies." Learned scientific groups have weighed in on the ethics of human gene editing technologies and we should be preparing our medical students and trainees to consider the ethical issues as they are likely to encounter these, and similar, technologies in their professional practice. Beyond patient care, what should be the role of physicians in stimulating and informing public discourse?

Regenerative medicine: This has the potential to improve the prognosis and quality of life for individuals with a multitude of chronic diseases and disabilities. Possibilities include using tissue engineering, reprogramming an individual's cells or inserting stem cells to replace or restore damaged or diseased organs and structures. Not only will regenerative medicine lead to an array of therapeutic possibilities, it could provide hope for many patients who now suffer from irreversible problems. How could something with such great potential raise ethical issues?

- One example is the ethical debate around the use of embryonic stem cells. While induced pluripotent stem cells (IPSC) may render this issue moot, it is likely that IPSC will not be optimal for all applications.
- It is likely that many if not all regenerative medicine interventions will be expensive. How will the decision be made as to who is a candidate? Should ability to pay enter into the decision-making? While this issue is not unique to regenerative medicine, as it arises with the introduction of many new disruptive therapies, it is sure to be a recurring ethical conundrum.
- Therapeutic applications of regenerative medicine will likely initially be for patients with the most serious of conditions such as spinal cord injury or amyotrophic lateral sclerosis. However, it may well be years or decades before unanticipated adverse complications are recognized. Will there be uncontrolled cell proliferation? Unexpected cell differentiation? As therapeutic indications are broadened, how will physicians strive to ensure that true informed consent is given, balancing the promise around the potential good with the understanding that there may be long-term unforeseen negative effects of cellular manipulation?

These are only some examples of potential ethical conundrums that will arise from progress being made in regenerative medicine.

Similarly, the ability to grow and harvest organs from animals such as pigs for transplant in humans—xenotransplantation—will soon be realized. While potentially alleviating the shortage of organs for transplant and thereby decreasing the morbidity and mortality of patients on long waitlists, there are individuals who due to religious beliefs might find such an organ unacceptable. What should be the physician's responsibility to inform and/or discuss alternatives?

The ethical issues physicians will face in the coming decades will increase in number and complexity. It is likely that for many, there will be little precedent to call upon. That said, it is an abrogation of our duty as medical educators if we do not prepare our medical students and residents to thoughtfully engage with their patients around ethical issues as well as to participate in the informal and formal public discussion. As our nation becomes ever more pluralistic and multicultural, understanding that the ethical and moral positions of one's patients may differ from one's own viewpoints is important. Furthermore, differentiating between legal and moral "right" and the ever evolving societal discourse on ethical issues will continue to complicate clinical decision-making. Didactic presentations, while useful for providing a common vocabulary and an introduction to general principles, are insufficient. Involving learners through case presentations and participation on ethics consultation services should be considered.

Potential learning enhancements

- Present a scenario that involves patients from a non-western background and ethical concerns arising around end of life and truth telling issues. For example, when the firstborn son requests that his elderly mother not be told she has cancer. Or, disagreements around the termination of life support. Discuss in small groups with learners representing the different sides of the conundrum.
- Explore the ethical and privacy issues that arise from sensor-based continuous monitoring.
- Discuss the ethical issues arising from the commercialization of patient databases. Consider referencing the Henrietta Lacks story and the HeLa cell line.

References

[1] Cruess SR, Cruess RL. Professionalism: a contract between medicine and society. JAMC 2000;162(5):668–9.
[2] Markel H. "I swear by Apollo"—on taking the Hippocratic oath. N Engl J Med 2004;350(20):2026–9.
[3] Parsa-Parsi RW. The revised declaration of Geneva a modern-day physician's pledge. JAMA 2017;318(20):1971–2.
[4] Committee on Quality of Healthcare in America, Institute of Medicine. Crossing the quality chasm: a new health system for the 21st century. Washington, DC: National Academies Press; 2001. p. 6.
[5] Tewarie B, Bailey V, Rebarbar M, Xu J. Unmet needs: hearing the challenges of chronic patients with artificial intelligence. NEJM Catalyst January 30, 2019. https://catalyst.nejm. org/unmet-needs-patient-forum-chronic-patients/?utm_campaign=editors-picks&utm_ source=hs_email&utm_medium=email&utm_content=69914587&_hsenc=p2AN- qtz-8HSl10GzbiwD5eExYD0onX6oiPX43qKHpzhFcxMW4naYqM9XSw0VlSdR05X- p9AmP3zJQPvlaiREefBM94-OkI9RrgiXA&_hsmi=69914587. (Accessed February 14, 2019).

[6] Spencer-Bonilla G, Quinones AR, Montori V, On behalf of the International Minimally Disruptive Medicine Workgroup. Assessing the burden of treatment. J Gen Intern Med 2017;32(10):1141–5.

[7] Grant RW, Adams AS, Bayless EA, Heisler M. Establishing visit priorities for complex patients: a summary of the literature and conceptual model to guide innovative interventions. Healthc (Amst) 2013;1:117–22.

[8] Tingley K. Trying to put a value on the doctor-patient relationship. N Y Times Mag May 5, 2016. https://www.nytimes.com/interactive/2018/05/16/magazine/health-issue-reinvention-of-primary-care-delivery.html.

[9] Barnard D. Vulnerability and trustworthiness. Polestars of professionalism in healthcare. Camb Q Healthc Ethics 2016;25:288–300.

[10] Pellegrino ED. Toward a reconstruction of medical morality. Am J Bioeth 2006;6(2): 65–71. https://doi.org/10.1080/15265160500508601.

[11] Murray E, Pollack L, White M, Lo B. Clinical decision-making: patient's preferences and experiences. Patient Educ Couns 2007;65:189–96.

[12] Schwartz LM, Woloshin S. Medical marketing in the United States, 1997–2016. JAMA 2019;321:80–96.

[13] Kelly T. Conflicts about conflict of interest. A comparison of performance-based and trustworthiness models in the context of detailing and gifts to physicians. Camb Q Healthc Ethics 2016;25:526–35. https://doi.org/10.1017/S0963180116000177.

[14] Pellegrino ED. Medical ethics in an era of bioethics: resetting the medical professions compass. Theor Med Bioeth 2012;33:21–4.

[15] Medical Professionalism Project. ABIM Foundation, ACP-ASIM Foundation, European Federation of Internal Medicine. Medical professionalism in the new millennium: a physician charter. Ann Intern Med 2002;136(3):243–6.

[16] Bauchner H, Fontanarosa PB, Flanagin A, Thornton J. Scientific misconduct and medical journals. JAMA 2018;320(19):1985–6.

[17] Hendriks F, Kienhues D, Bromme R. Measuring lay people's trust in experts in a digital age: the Muenster Epistemic Trustworthiness Inventory (METI). PLoS One 2015;10(10): e0139309. https://doi.org/10.1371/journal.pone.0139309.

[18] AMA Council on Ethical and Judicial Affairs. Code of medical ethics of the American Medical Association. American Medical Association; 2016. ISBN: 978-1-62202-553-4.

[19] Jenkins S, Draper H. Care, monitoring, and companionship: views on care robots from older people and their carers. Int J Soc Robot 2015;7:673–83. https://doi.org/10.1007/s12369-015-0322-y.

[20] Draper H, Sorell T. ACCOMPANY Deliverable 6.6 a tentative proposal for an ethical framework. European Commission, FP7-ICT-2011-07, Framework Programe ICT Call 7 – Objective 5.4 for Aging and Wellbeing; 2014.

[21] Sharkey A, Sharkey N. Granny and the robots: ethical issues in robot care for the elderly. Ethics Inf Technol 2012;14:27–40. https://doi.org/10.1007/s10676-010-9234-6.

[22] Chen S, Pan ZX, Zhu HI, Wang Q, Yang J, Lei Y, et al. Development of a computer-aided tool for the pattern recognition of facial features in diagnosing turner syndrome: comparison of diagnostic accuracy with clinical workers. Sci Rep 2018;8:9317. https://doi.org/10.1038/s41598-018-27586-9.

[23] Martinez-Martin N. What are important ethical implications of using facial recognition technology in health care? AMA J Ethics 2019;21(2):E180–7.

[24] Chaet D, Clearfield R, Sabin JE, Skimming K, On behalf of the Council on Ethical and Judicial Affairs American Medical Association. Ethical practice in telehealth and telemedicine. J Gen Intern Med 2017;32(10):1136–40. https://doi.org/10.1007/s11606-017-4082-2.

Additional resources

- Bulger RJ, McGovern JP, editors. Physician and philosopher: the Philosophical Foundation of Medicine: essays by Dr. Edmund Pellegrino. Charlottesville, VA: Carden Jennings Publishing; 2001. ISBN: 1-891524-09-7.
- DuBois JM, Anderson EA, Chibnall JT, Diakov L, Doukas DJ, Holmboe ES, et al. Preventing egregious ethical violations in medical practice. J Med Regul 2018; 104(4):23–31.
- Nicholson Price II W, Cohen IG. Privacy in the age of medical big data. Nat Med 2019;25:37–43.
- Byyny RL, Papadakis MA, Paauw DS, Pfiel S. Medical professionalism best practices: professionalism in the modern era; 2017. Alpha Omega Alpha Honor Medical Society. ISBN: 1532365160, 9781532365164.

Communication

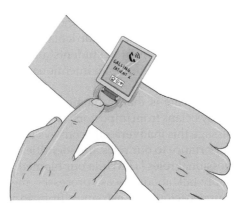

Communication has always been an essential part of a physician's professional role. Indeed, one's success as a clinician is dependent upon the ability to communicate effectively. An integral part of medical education is learning how to communicate with patients and their families, how to convey bad news, how to propose techniques to enhance behavior change, how to elicit informed consent, and how to summarize a patient's history both orally and in text. However, beyond the immediacy of clinical care focused communication, physicians also inform broader audiences regarding medical and public health issues, communicate with payers, advocate for their patients, and influence regulatory agencies and governmental policy makers. These audiences and more are regularly the focus of physician communications.

"I think it's not going to be that hard to be a technically excellent physician… that's a learned skill and we have so many tools that are making it easier. I think the great physicians are going to be people that are able to communicate properly …have good rapport with their patients, their patients appreciate them."

Stephen Papadopoulos

Implementing Biomedical Innovations into Health, Education, and Practice
https://doi.org/10.1016/B978-0-12-819620-5.00004-7

Medical and biomedical sciences lexicon

At its most basic level, understanding the language of medicine is necessary to not only communicate with other clinicians and healthcare professionals but also to understand new developments and advances. Ranging from anatomy to -omics to pharmacotherapeutics to technology, the vocabulary of medicine is extensive. Learning the language of the profession to understand what is being communicated, whether it be through scientific communications, clinical updates, or with professional colleagues, will remain an important component of medical education.

As noted in the discussion in Chapter 14 regarding premedical requirements, I propose that an understanding of the medical and biomedical science vocabulary be a requirement for matriculation. Rather than the oftentimes happenstance development of students' understanding of medical and scientific terms, this should be an intentional part of their earliest medical education.

The language of medicine is that of biomedical science, as science is what distinguishes physicians from impostors. However, it is worth noting that the implicit message this inadvertently conveys is that biomedical science is of greater importance to our roles and identities as physicians than the healer and professional roles [1]. While our biomedical vocabulary allows us to comfortably talk about disease processes, mechanisms and interventions, does it unconsciously constrain us from addressing suffering, loss, mourning, spiritual and transcendent dimensions? Conversations about curing a disease are accommodated but the many facets of healing, especially when cure is no longer possible, may be constrained. Medical educators must thoughtfully consider whether or not our very language skews the early identity formation of our students away from their roles as healers.

Patient and family communications

One of the challenges lies in imagining what communication will be like in midcentury. When one considers the rapidity of change in the late 20th and early 21st century, going from landline and pay telephones to the ubiquitous presence of smart phones, from letters and faxes to email and texting, from television with fixed programming to a virtually infinite variety of programming on demand, from face-to-face meetings to videoconferencing, it is hard to envision the technological capabilities that will be available for communication. The possibility of facial recognition coupled to an individual's genetic, environmental, and even behavioral database with near instantaneous algorithmic identification of pertinent history and risks available to the physician on their heads up display is not so farfetched. What would such a scenario mean for the patient physician interaction, whether synchronous or asynchronous in time and space?

Regardless of the advances in technology and proliferation of diagnostic, coaching, and similar apps, the importance of human to human communication will remain. While it likely will be acceptable to most individuals to be informed by a computer of a diagnosis of rhinovirus infection, learning of a terminal diagnosis or one that portends significant lifestyle alterations and morbidity will likely still be best accomplished through human to human interaction.

> "I'm much more of a humanist here and I think people want to be touched and want to be talked to … the dispensing of care in medicine is still personalized and still one person at a time… people don't want to interact with the kiosk… they (the patient) want to hear it from someone and some comfort that goes with telling them what's going on, whether that's good news or bad news."
>
> *Lawrence Correy*

But it won't be only the major clinical events that will require human to human communication. It will also be to help patients and their families make sense of the sometimes conflicting information available from a myriad of sources and make the decisions that are best for them. This speaks to physicians' roles as a trustworthy and honest broker.

> "… Why I think this is particularly relevant to the future of medicine and how we train, is that there are going to be so many people navigating information that is available to them or even care choices and decisions that they would be making on their own in terms of getting their own testing … we (physicians) are going to be put much more in the role of enabling people to make good decisions. Let me deconstruct your thinking on this, understand how you got to this place, and with your permission, criticize that to see if that's the right place, or you should be at a different place because these things that you have chosen to order, the people you want to see, have all been shaped by these assumptions. So how do we start to get to evidence-based narratives from the patient and move more from their interpretive side while not dishonoring that?"
>
> *Joseph Kolars*

While the admonition to avoid medical and scientific jargon when speaking with patients and their families is well known, less appreciated is variable interpretation of commonly used words. A recent study compared seriously ill patients' and physicians' interpretations of the word "treatable." From the physician's perspective, saying the patient's illness was treatable meant that there was an action or intervention available but it did not imply cure or a better prognosis. From the patient's perspective, being told their disease was treatable was viewed as conveying a positive message and good news that their quality of life would improve with the hope for increased length of life and even cure [2]. As diagnostic and therapeutic options proliferate, medical students and residents need to learn how best to explain in understandable terms the often complex scientific principles. And, as importantly, how to check to make certain that patients and their families interpret the implications of what is being said correctly.

Another area of potential confusion is in communicating risks and probabilities. Information about complications or benefit is presented to patients without clarifying whether absolute risk or relative risk is being discussed. For example, if the absolute risk of developing a particular disease over one's lifetime is 1 in 5000 but through yearly screening or another intervention the absolute risk can be reduced to 1 in 10,000 then the relative risk of developing the disease would be decreased by 50% through the intervention. Many individuals when presented with an option to reduce their risk of developing a disease by 50% would be more predisposed to pursue the screening procedure or intervention as compared to hearing that the risk of them developing disease decreased from 1 in 5000 to 1 in 10,000 [3]. Too frequently physicians and other clinical personnel do not clarify to patients or their surrogates whether absolute or relative risk is being discussed. When the benefits of an intervention are being extolled, relative risk is often used, whereas when complications are being discussed, the absolute risk is used as it has the effect of downplaying the likelihood of an adverse outcome.

Not surprisingly, studies have shown that patients' personal medical decision-making is affected by multiple biases and heuristics [4]. Physicians and all clinicians need to appreciate that to truly make informed decisions, patients, their families and/or caregivers need to understand what is at stake. Using precise language and ensuring that patients fully grasp all of the implications of what is being discussed cannot be overemphasized.

Beyond the fundamental principles, advances in technology have provided an ever-growing list of communication channels. As noted, it is impossible to predict the many modalities that physicians will utilize mid-century. However, certain broad categories must be considered.

In person communication

In person communication with the patient, their family and/or caregivers has been the primary mode of interaction for centuries. While the emphasis of medical student and trainee instruction has expanded from inquiring about symptoms and explaining diagnostic and therapeutic interventions to include developing rapport, questioning strategies to elicit the patient's perspective and understanding, nonverbal communication, motivational interviewing, and more [5, 6], the focus of most clinician training remains on in person communication.

An essential part of in person communication is recognizing the role of culture in defining appropriate communication morays. As the population of the United States becomes ever more heterogeneous, the necessity of understanding the cultural background of one's patients and caregivers will only increase. One simple example, where I grew up in rural Minnesota I was taught that a firm handshake and eye to eye contact is indicative of respect and trustworthiness. A limp handshake or not looking the person in the eye when speaking to them was viewed with suspicion. It also was impolite to invade another person's personal space which was at least 2–3 ft, even for good friends. Other cultures have very different communication norms. Looking a person in the eye when speaking with them might be interpreted as aggressive or condescending. Having a conversation with a 2–3 ft personal space is being standoffish or disrespectful. While in person communication provides a richness of exchange that is not readily accomplished with other modalities it also can lead to misunderstanding if cultural differences are not recognized.

Part of the power of in person communication is the nonverbal component of the interchange. Often unconsciously, empathy, understanding, agreement or disagreement is communicated nonverbally. While many aspects of nonverbal communication are universal, it too has nuances that are learned and are culturally situated. Due to lack of knowledge regarding cultural norms, insulting or even offensive actions may be committed. Showing the soles of your shoes, touching or offering a person something with your left hand, and gestures such as a thumbs-up are all examples of what most Americans consider innocuous actions that are deemed impolite to offensive in some cultures.

Expanding existing communication training to include explicit instruction in culturally determined communication norms is important. Simulations are powerful learning modalities to develop medical student's comfort with and understanding of different communication styles. To ensure optimal clinical care, future physicians will need sophisticated, culturally appropriate, in person communication skills to engage with their patients, families and caregivers.

Asynchronous communication

But the focus can no longer be solely on in person dialog as the reality is that asynchronous modes in space and time will become ubiquitous means of communication between clinicians and patients, their families and caregivers.

"As a health system, we are developing alternative specialty care delivery models – telehealth, home hospitalization, etc – with the aim of bringing care to the patient. At some point in the future, we expect that the number of patients cared for in this way will exceed those cared for in an office setting."

Elizabeth Nabel

Whether it is delivering care in underserved areas, providing subspecialty consultations in geographical locations without such specialists, or remaining in contact with patients outside the clinic, asynchronous communication is being used in numerous settings and across multiple applications. Already, electronic interactions account for a majority of "visits" in some systems.

Developing a "webside manner" increasingly is recognized as central to medical education. As it requires a different skill set than in person communication, specific programs are being developed to teach students how to conduct clinical examinations via videoconferencing, monitor data from wearable devices, and understand how to effectively engage patients [7]. Although curricular offerings are still in the early stages of development, the fact that in many schools, instruction in asynchronous communication is seen as important for medical student education speaks to its growing importance.

Let us consider some specifics.

Videoconferencing technology, while asynchronous in space, is a reasonable proxy for in person discussion. It allows the listener to infer meaning beyond the specific words through voice intonation, some body language, and nonverbal communication. That said, clinicians will need to recognize technologic constraints such as limited field of vision, to minimize the likelihood of miscommunication, especially when multiple individuals are involved in the conversation. In addition, while the convenience factor of videoconferencing is notable, there are a multitude of factors that enhance the therapeutic value of a physician-patient relationship that are not achievable with technology. An obvious example is the healing power of culturally appropriate "touch" that by definition is lost with asynchronous communication.

"… we don't teach them (medical students and residents) how to be good telemedicine doctors since that's what most of them will be predominantly. … but also we're not using these tools, the tools that are all ready to be incorporated into the daily practice of medicine …."

Eric Topol

More challenging is communication through social media or messaging for tasks beyond straightforward information exchange. In addition, if only text is employed, a message devoid of voice intonation and body language increases the likelihood of misinterpretation. Eliminating the opportunity for real time clarification and questions further limits the instances in which asynchronous in time and space communication is clinically optimal.

Although the technology that enables asynchronous communication is widespread, little work has been done in the clinical realm to identify best uses for specific platforms and to study how best to ensure accurate interpretation of the message and maximize its utility while recognizing its limitations. As asynchronous modes of communication will become ever more common between patients and clinicians, research is needed to identify cultural differences and avoid unintentional faux pas that can undermine patient confidence and trust. To date, the enthusiastic early adopters tend to minimize the weaknesses of the technology and the naysayers focus on the disadvantages. While the convenience is unquestioned, thoughtful studies of the various platforms, ranging from videoconferencing to email to instant messaging are necessary. As communication is essential for the patient physician relationship, ensuring mutual understanding should not be left to chance and developing the appropriate skills in our students and residents requires explicit attention.

Regardless of the technology employed to communicate with patients, their surrogates and caregivers, we know what patients want. It is unlikely that these desires will change dramatically despite changes in communication modalities.

Patients want their clinician to:

- listen to them,
- tell the full truth about their diagnosis even though it may be uncomfortable or unpleasant,
- tell about the risks associated with each therapeutic option,
- explain how the options may impact their quality of life, help them understand how much each option will cost them and their family, and
- have the clinician understand their goals and concerns regarding the various options [8].

Expanding the communication skills of the physician to include facility with the full panoply of options will be a challenge for medical educators. Technology will continue to increase the means of communication but will not necessarily result in enhanced understanding or strengthen the therapeutic value of the patient clinician interaction. Just as clinicians are taught the appropriate use of diagnostic and therapeutic tools, so too the explosion of communication modalities requires explicit instruction and learning in order to be used properly.

Public communication

Beyond patients, their families and caregivers, physicians are often turned to as knowledgeable experts to provide understandable explanations of scientific findings, public health recommendations, and issues related to health and disease in the news cycle. With the continued proliferation of "information" sources, some of which are of dubious value, the importance of the physician's role as a source of valid information or "truth" likely will grow. To effectively communicate to a broad public audience requires developing an appreciation for the level of understanding of one's listeners and being able to "connect" with them at an intellectual and, very importantly, an emotional level. The adage "data is important, but stories are memorable" is especially applicable to public speaking. Frequently physicians fall into the trap of relying on our biomedical language and trying to be comprehensive in our statements or explanations. Explicitly equipping our students and trainees to take what sometimes are complex scientific and ethical issues, assess their audience and develop readily understandable and compelling narratives will require medical educators to expand the focus of communication curricula.

"the doctor in my view should be one of the bulwarks against medievalism and ignorance as it affects medical care… and the medical school has to be a bastion for that, it may not be a job people want to do but there's nobody else who will do this. So we can't live in a world where people say vaccines are harmful and then get away with that. I know we're not supposed to be judgmental, I know we're not supposed to be certain, but vaccines changed a huge amount of medicine. We don't have polio anymore. Most doctors don't see diphtheria, at least didn't, but now they're seeing it again."

Samuel Broder

As it is likely that that their professional careers will also require communicating with regulators, government officials developing policy, community leaders and activists, and the media, learning how to engage with widely varying audiences and craft a consistent message in language and style tailored to those being addressed should be included in the communication training of all students and residents.

Professional communication

The importance of communication among professionals caring for patients is self-evident. For generations, medical students were given the general outlines or templates for "oral presentations," "write ups" and "clinical notes" and then were expected to develop their professional communication skills while on their clinical rotations. But, changes in medical practice and the introduction of technology necessitated increased attention to professional communication. Let us consider this further.

One thing that has not changed is the fact that words matter. How a person is characterized subconsciously or consciously frames the clinician's thinking. Describing a person as the "obese diabetic" in room 403 or "Mrs. Smith who has diabetes" in room 403 evokes different mental images. The words used to describe a patient may evoke negative conscious or unconscious biases, dehumanize and devalue the person, or encourage engagement, care and recognition of another human being with a life history and value beyond the disease that brings them to the clinical system. This fundamental truth must continue to be taught and role modeled in all professional communications regarding patients.

Professional communication is an example of the unintended consequences of technologic applications. The limitations of current electronic health records (EHR) are well known. Promulgation of erroneous data often times by using copy and paste rather than actual updating of progress, pulldown menus that facilitate efficiency but might not accurately capture important history or physical findings, the use of templates which inadvertently lead to the patient fitting the standard template rather than the communication being tailored to that specific patient, and the unprofessional but recurring practice of "documenting" a history that was not actually elicited or something that was not actually done—for example foot inspection, a comprehensive review of systems, or review of medications and allergies. The potential for harm is obvious. While such documentation issues are not new, the introduction of EHRs has exacerbated the problem and is an example of unanticipated, but potentially catastrophic, consequences of a technologic advance. While seeming self-evident, the paramount importance of accuracy and truthfulness must be emphasized.

For physicians, the format of oral presentations, admission write ups, progress notes, consults and discharge summaries that were utilized for decades allowed listeners or readers to anticipate when specific information would be provided. This facilitated understanding not only of the patient's course but also the clinical logic underlying diagnostic and therapeutic decisions. Another recurring complaint is that EHR based consultations, discharge summaries, and notes to referring physicians are replete with pages of test reports but do not allow the reader to follow the patient's progression or the logic supporting the clinical thinking. While it is likely that "written" records will evolve to maximize the efficiency and quality of information technology based communication modalities [9], it is apparent that professional communications must evolve to fulfill their diagnostic and therapeutic intent and not simply meet billing requirements. Just continuing to teach the traditional formats does not take advantage of the technology's potential and the reality is that professional communication is too frequently impeded rather than enhanced.

Studies have shown that communication among the many professionals providing care for a given patient is of paramount importance for patient safety as well as for appropriate care. The introduction of duty hour regulatory requirements and new models of inpatient care employing hospitalists and nocturnalists have increased the number of inpatient "handoffs." In turn, this has multiplied the opportunities for harm due to inadequate or incomplete conveyance of patient information. An ongoing focus of investigation is identifying optimal information handoff tools. As this will vary from discipline to discipline, it is an appropriate aspect of resident education programs.

In addition, as the scope of clinical disciplines involved in the care team expands, there is a need for clarity of language. Is the individual being cared for a patient or a client and are there differences in one's responsibilities when interacting with a patient as compared to a client? Not only are there potentially important differences in foundational assumptions, there may also be differences in the meaning of commonly used words and phrases. What is normal, unremarkable, and appropriate for age? Explicit attention to developing a common understanding of what the different professionals in the team are conveying is important. Potentially, team-based interprofessional education will provide the impetus to develop a common understanding and clinical language but our future physicians and other professionals need to recognize this potential for inadvertent miscommunication.

These are but a few examples of how professional communication must continue to evolve in concert with our evolving clinical practices and why attention to its importance is needed.

Consequential decisions—Trustworthiness and communication

Although I have chosen to consider trustworthiness and communication separately this is an artificial distinction. They are inexorably linked. This is especially true when faced with consequential decisions. Many patients, their families and caregivers will face weighty decisions with life-and-death implications. The decision as to which therapeutic option to select for newly diagnosed cancer, whether to terminate chemotherapy and enter palliative care, to discontinue ventilatory assistance, to recognize that hospice care is an appropriate next step—when patients, their families and caregivers are faced with momentous decisions such as these, trust and clarity in communication are essential. In what oftentimes are emotionally charged interactions, the clinician must strive to convey realistic expectations and respect the wishes of the patient—fulfilling their healer role—while also considering societal expectations. To do so requires that the patient's current priorities are elicited, as these can change over time as disability and other factors arise, and that the patient clearly understands the implications of their decisions.

As discussed in patient or person centeredness, the paternalistic approach to making decisions for patients has rightly been criticized. However, the extreme alternative of simply providing patients, their families and caregivers all potentially relevant data is similarly inappropriate. If the physician has invested the time to understand the patient's priorities, it is appropriate to guide the patient through the decision-making process. For example, the statistics that we as professionals provide are essentially meaningless to an individual. We quote statistics such as a 10% chance of complications, however, for the patient it is 0% or 100%. We may use terms with which the patient is familiar but doesn't grasp the full import of the term, such as incontinence as a long-term sequela of a prostatectomy. We do not say, "you might need to wear a diaper." Moreover, studies have repeatedly shown that individuals faced with a major life altering event, such as being told they or their child have cancer, often do not "hear" the rest of the conversation as the emotion is overwhelming. For patients facing major decisions, it is appropriate for the physician, who has the knowledge and the objectivity, to guide the individual through their medical options and decision-making. To simply present facts regarding alternatives, absent consideration of patient preferences, is an abrogation of a physician's responsibilities to their patient.

"…as a physician one of my greatest medical roles is to help people understand what a doctor is telling them, what a medical team is telling them, or interpreting what this illness might be or how would this therapy work."

Ellen Sheets

There are many factors that may further complicate the situation. The individual may be in an intensive care unit and the clinician(s) caring for the patient have not had a long-standing relationship with the patient. The ethnic, racial, or religious background of the patient and their family may differ from that of the clinicians resulting in conflicting assumptions and understandings. Family members who have had little interaction with the patient may have unrealistic expectations for stabilization and recovery. Likely, the financial consequences of care decisions will increase and be borne by the individual patient and their family—a factor that might not be verbalized or even acknowledged.

As the potential for misunderstandings and disagreements is high, the essential requirements for a satisfactory outcome are clarity of communication and a level of trust that dispels any concerns that the clinician is not acting in the best interest of the patient. Regardless of the multitude of changes likely to occur in the practice of clinical medicine, the central role of the clinician in the discussions that precede such consequential decisions will remain. The human to human interaction and "touch" will remain essential. That said, the complexity of these interactions will increase and there should be a commensurate increase in the training of medical students and residents in how to initiate and carry on these difficult conversations.

Communication, whether with patients, professional colleagues, or the multitude of other stakeholders is a constant in clinical medicine. It is high-stakes with the potential for serious harm if incomplete or misunderstood. While our attention may be on the tools used for communication, the multitude of purposes it serves in clinical medicine should remain the focus. Technologic platforms will continue to enhance the potential for communication but also pose often times unanticipated challenges. As communication modalities evolve, the task for educators will be to ensure that our students and residents continue to understand the importance of accurate information exchange, the development of mutual understanding across multiple contexts and purposes, and the ramifications, both positive and negative, of how clinicians communicate.

Potential learning enhancements

- Through simulated patient exercises or videotaped vignettes expose students to communication norms from cultures other than their own. Have them reflect on their degree of comfort or discomfort and why.
- Have faculty trained in cross-cultural communication or people in the community from differing cultures present and demonstrate to the students the communication norms that are specific to the culture of patients with whom they will be interacting.
- Have students assume the role of physician and communicate the therapeutic or survival benefits of a new intervention in multiple ways. Have the students who are observing rate the persuasiveness of the different options and reflect on why they gave the rating.
- Invite a guest speaker who is in advertising or marketing to speak with the students regarding the use of language and presentation to persuade.
- Present the risk and probability for an outcome in different ways, alternating between relative and absolute risk and in a positive or negative statement, and discuss how it "felt."
- Have students review a written case summary and then present the patient orally. Review and critique video recordings of their presentations and provide an "expert" presentation example for comparison.
- Simulate an asynchronous, such as email, patient initiated inquiry asking the learners to respond. Review and critique the responses.

References

[1] Sternszus R, Regehr G. When I say … healing. Med Educ 2018;52:148–9. https://doi.org/10.1111/medu.13413.

[2] Batten JN, Kruse KE, Kraft SA, Fishbeyn B, Magnus DC. What does the word "treatable" mean? Implications for communication and decision-making in critical illness. Crit Care Med 2019;47(3):369–76.

[3] Andrade C. Understanding relative risk, odds ratio, and related terms: as simple as it can get. J Clin Psychiatry 2015;76(7):e857–61.

[4] Blumenthal-Barby JS, Krieger H. Cognitive biases and heuristics in medical decision-making: a critical review using a systematic search strategy. Med Decis Making 2015;35:539–57.

[5] Smith S, Hanson JL, Tewksbury LR, Christy C, Talib NJ, Harris MA, et al. Teaching patient communication skills to medical students: a review of randomized controlled trials. Eval Health Prof 2007;30(1):3–21.

[6] Keifenheim KE, Teufel M, Ip J, Speiser N, Leehr EJ, Zipfel S, et al. Teaching history taking to medical students: a systematic review. BMC Med Educ 2015;15(1):159.

[7] Warshaw R. From bedside to webside: future doctors learn how to practice remotely. AAMC News. Medical education. April 24, 2018. https://news.aamc.org/medical-education/article/future-doctors-learn-practice-remotely/. (Accessed April 25, 2018).

[8] Alston C, Paget L, Halvorson G, Novelli B, Guest J, McCabe P, et al. Communicating with patients on healthcare evidence [Discussion paper]. Washington, DC: Institute of Medicine; 2012.

[9] Skeff KM. Reassessing the HPI: the chronology of present illness (CPI). J Gen Intern Med 2013;29(1):13–5.

Economics

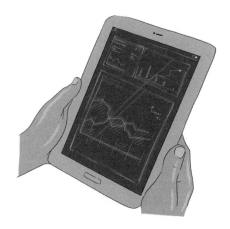

For decades, medical education ignored economic realities. When economics was considered, often near the completion of residency, it was in the context of choosing a practice and focused on the business aspects of medicine. This has changed and likely will continue to assume greater importance as the expenditures on "healthcare" as a percentage of national gross domestic product (GDP) continue to rise. The conversation has now grown to include what is the value of the expenditures to our society as a whole given our poor rankings on many Organization for Economic Cooperation and Development (OECD) health outcomes. While administrative waste, defensive medicine, unnecessary testing, fraud, pharmaceutical pricing, lack of pricing transparency, physician compensation, hospital charges and the like have all been implicated in the ever-increasing cost of medical care, the reality is that physicians play a considerable role in determining expenditures directly or indirectly through their ordering and prescribing patterns. As such, it is imperative that students and trainees develop a basic understanding of the economics of clinical care and the physician's role therein.

Implementing Biomedical Innovations into Health, Education, and Practice
https://doi.org/10.1016/B978-0-12-819620-5.00005-9

> "Medicine has changed you have to understand, I mean shame on us if our medical students don't understand the balance sheet, income statement, and cash flow. It's not that hard but we need those real foundational building blocks understood because it's used in everything we do now."
>
> *Stephen Papadopoulos*

As patients and their families are required to directly bear greater financial burden for their medical care, the microeconomic implications of clinical decisions will be an increasingly important factor. Physicians will be expected to consider the trade-offs between expected improvement in health or modification of morbidity and disability versus cost for different clinical options they recommend. In addition to value judgments for their individual patient, they will be expected to consider the financial impact to their organization. It is likely that physicians' clinical decision-making, in partnership with their patients, will require consideration of the microeconomic implications of a recommended diagnostic evaluation or course of therapy. This is already starting to happen.

Regardless of the specifics of the payment mechanism, it is anticipated that these financial exigencies will continue. Educating medical students and residents as to the economic implications of their decisions and understanding how cost-benefit analyses, utilities, and similar analyses will help in their decision-making will be a fundamental aspect of the clinical practice.

While most students and residents will predominantly focus on the microeconomic aspects of medicine, they will also need at least a passing understanding of the macroeconomics at play. As the imperative for cost control of medical care grows, there will be responses from payers and regulators, from concerned outside watchdog groups, and from within the profession itself. As physicians are estimated to account for approximately two-thirds of medical costs and unnecessary services are estimated to contribute over $200 billion to the estimated more than trillion dollar annual "healthcare waste," their informed engagement is important. Understanding the value of care and eliminating no value or low value care is a first step. While intuitively obvious and seemingly straightforward, for most medical services the value depends upon the clinical situation. While consensus regarding the components to include when defining "value" has been elusive there are commonalities that should be part of the knowledge base of all students and residents.

$$\left[\text{Value} = \text{quality} / \text{cost or } V = (\text{quality} + \text{service} / \text{direct and indirect cost}) \times \text{appropriateness}\right]$$

- One is *quality*. However quality depends upon outcomes and for many medical services and interventions robust outcome metrics are lacking
- Another is *cost*. However is this charges, what a payer has contractually agreed to reimburse, or production cost? Also, does cost only include direct or should it include indirect components such as opportunity costs?
- Should the patient experience as defined by "service" be included in the numerator? If so, what are the metrics that should be used to define the patient experience?
- And finally, the whole concept of *appropriateness* must be considered.

As the sophistication of analyses grows, it will be important that physicians understand concepts such as quality adjusted life years, incremental cost-effectiveness ratios, and how substantial changes in price will potentially change the assessment of an intervention from cost ineffective to cost-effective. It is likely that their personal compensation and their medical organizations reimbursements will be tied to assessments of the value of care provided. Understanding the nuances and complexities of these computations will be essential.

Although there is acknowledgment that the assessment of value should include non-monetary factors from the patient's perspective, too frequently this is not the case. Most discussions defer to financial analyses and outcomes that reflect payer and clinician perspectives but not the patient's perspective. Indeed, what are deemed low value services may be perceived as having value in the patient's eyes. Eliminating these "low value" services may be interpreted as something being taken away from them rather than an opportunity to provide higher value care at the same or reduced cost. This conundrum highlights the importance of acting as an honest broker, having already established trustworthiness in the eyes of one's patients.

"Our ability to invest more money, not only in our own health care delivery system here, but globally, we shouldn't count on that. Which means that the investment needs to be on value, on improving the value, and I think that's the driver of healthcare.…I cannot imagine 20 to 30 years from now that the status quo is sustainable in any way shape or form… The public is just going to say, 'Stop! We are done investing. So what are you doing to increase their value for the revenue we're already contributing?'"

Rajesh Mangrulkar

Resource allocations and payment decisions will be made with or without the involvement of physicians. In my opinion, physicians need to be involved in the discourse and decision-making. Alternatively, payers and regulators will make decisions and policies with potentially negative ramifications for clinical care. It behooves students and trainees to be familiar with the macroeconomic basis of clinical medicine and to be able to follow, participate in, and for some, ultimately lead the discussions.

Potential learning enhancements

- Provide the students an anonymized "routine" hospitalization. Have them estimate what the cost would be in total and to the patient if paid by a commercial carrier; if paid by Medicare; or out of pocket if uninsured. Compare that to the actual bill in each circumstance.
- Have the learners access two or more area hospitals' charge masters and compare prices across institutions. Discuss potential explanations for the observed variability.

Diagnosis and clinical reasoning

The correct disease diagnosis is the cornerstone of clinical medicine as it has been for millennia. Absent the correct identification of a disease, by definition, physicians cannot practice quality medicine, resources are wasted and important skills such as patient centeredness and communication are for naught. Arguably, when the clinician's ability to intervene in order to cure or arrest the progression of disease is limited, the consequences of a missed diagnosis are less than when the appropriate treatment might return a patient to health. But even in such a circumstance, the suffering and potential morbidity associated with misguided therapy must be considered. As efficacious therapeutic options have multiplied and what were previously thought to be untreatable diseases have become treatable, the importance of accurate disease diagnosis has increased.

> "… disease is just a labeling system and the labeling is a short way of communicating what you think it is and diagnosis is a probability that this person meets this label."
>
> *K. Ranga Rama Krishnan*

Importantly, medical educators must imbue our students and residents with the mindset that the diagnosis of a disease is not finalized until the patient's problem is defined to the best of our scientific understanding. Making a diagnosis is not simply attaching a label to a set of data points that include the patient's presentation, biochemical and imaging studies,

Implementing Biomedical Innovations into Health, Education, and Practice
https://doi.org/10.1016/B978-0-12-819620-5.00006-0

but should, if possible, extend to defining the pathway(s) and pathogens, toxins, and other factors responsible for the disease. Just as a diagnosis of breast cancer is incomplete without further characterization of the disease as to receptor status and genetics for the purpose of guiding therapy, I anticipate that similar biomarker and cellular pathway definitions will become the norm for many diseases.

This will pose multiple challenges for medical educators. As our understanding of health and disease continues to expand, medical educators must provide their students and residents appropriate mental models, sometimes termed cognitive frameworks or schemas. These will help to incorporate new understandings of mechanistic pathways and expand their understanding of the development of disease, including earlier identification based on presymptomatic disease markers, as optimal therapeutic interventions are dependent upon accurate identification and characterization of the disease. A second challenge will be the re-classification of many diseases.

Let us consider an example. The constellation of symptoms that we associate with diabetes mellitus has been recognized for millennia dating back to ancient Egypt and India. But it was not until the late 1700s that it was recognized that individuals with diabetes had a sugar, subsequently identified as glucose, in their blood that was also excreted in the urine suggesting that it was a systemic disease. A major breakthrough was the 1921 isolation and identification of insulin by Banting and Best, leading to the development of effective therapy for many patients. Over the ensuing half-century, it became clear that while diabetes was characterized by elevated blood glucose levels there were differing pathophysiologic mechanisms and importantly that different therapeutic approaches are indicated. This led the National Diabetes Data Group to develop in 1979 a new diabetes classification system that included insulin-dependent or type 1 diabetes (distinguished by an absolute deficiency of insulin), non-insulin-dependent or type 2 diabetes, gestational diabetes, and diabetes associated with other syndromes or conditions. We now know there are dozens of etiologies that lead to clinical diabetes mellitus [1]. It is likely that as different underlying mechanisms are identified, new therapeutic approaches will be developed.

Reflecting on this brief history, for many years, physicians were taught how to recognize and name the constellation of symptoms as diabetes but there was no effective therapy. So even if they misdiagnosed the patient, the harm was minimal. Making the correct diagnosis became far more important with the introduction of insulin as effective therapy. This remained the state-of-the-art for several decades. Now however, we know that what once was called diabetes mellitus is actually multiple different diseases with different therapeutic options. Physicians must not only recognize the constellation of symptoms associated with diabetes mellitus but also obtain the biochemical studies to appropriately classify it as differing treatment approaches are indicated. While this is the current state of disease diagnosis, I anticipate that mechanistic classifications will further refine our diagnoses as additional therapeutic approaches are developed.

From an educational perspective, for much of the 20th century making the diagnosis of diabetes mellitus carried with it the therapeutic imperative for insulin and the implied mechanism of insulin insufficiency. As understanding grew, the response was to develop new classifications of "diabetes." The pace of change was slow enough that elderly individuals will continue to rise and those aged 80 and above will be going from a single diagnosis of diabetes to four subgroups was readily accommodated in our medical education system. However, can we similarly accommodate dozens, or even hundreds, of subclassifications? Moreover, in the future, new disease naming conventions likely will be developed. Are the mental models we impart to our trainees flexible and robust enough to incorporate such developments in real time? To impart the models of disease necessary to accommodate the rapidly expanding scientific knowledge requires an explicit effort. As shown in Fig. 6.1, the model we teach must

MODEL OF DISEASE

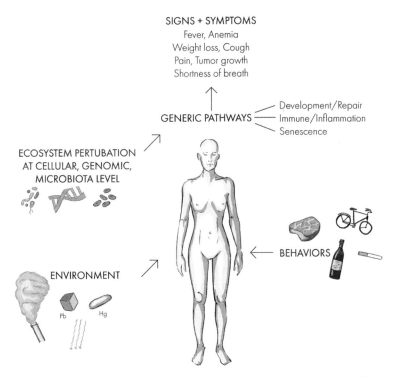

FIG. 6.1 Model depicting how disruption of an individual's ecosystem through external or intrinsic factors may manifest in signs and symptoms identified as disease.

have at its foundation the involved cellular pathways and mechanisms; include the triggering or predisposing factors; show the "generic" physiologic responses such as inflammation, oxidative stress, regeneration and repair, development and senescence that are activated; and the resulting signs and symptoms that the patient will likely manifest.

While this is an example of how disease classification likely will evolve, let us now consider the human factors and our current understanding. Unfortunately, the reality is that diagnostic errors are common and result in missed opportunities for early treatment, morbidity and death. A recent report from the National Academy of Medicine estimated that yearly 5% of adult patients seeking outpatient care experience a diagnostic error, that diagnostic errors account for up to 17% of adverse in-hospital events and contribute to approximately 10% of patient deaths. Further, that most individuals will experience at least one diagnostic error during their life [2]. These statistics highlight the need for enhanced attention in our educational and training programs to the development of diagnostic reasoning processes, judgment and decision-making skills of our students and residents.

There are multiple factors that contribute to diagnostic complexity and misdiagnoses. One is that while the human body is exposed to a multitude of insults leading to disease, we are relatively limited in our response options as reflected in signs and symptoms and physiologic perturbations. Indeed, while there are a few hundred potential presenting signs and symptoms, it is estimated that there are over 10,000 different diseases. As scientific advances have shown, diagnoses such as breast cancer and diabetes mellitus, the examples we considered, at one time thought to be singular entities, are actually multiple different diseases. When considered at the mechanistic level, the number of discrete disease entities will only increase. Adding to the complexity, some entities such as pulmonary emboli are highly variable in their presenting signs and symptoms and may be confused with other diseases that have similar presentations.

To improve the diagnostic process, an understanding of the reasons for diagnostic errors may be helpful. Studies have shown that many, if not most, diagnostic errors are due to the clinician's reasoning process. In a study of 583 physician-reported diagnostic errors, failure or delay in considering the diagnosis was the most common cause of diagnostic failures (19%), followed by failure or delay in ordering needed tests (11%), and erroneous laboratory or radiology interpretations of tests (11%) as assessed by the physician respondents. Thirty-one percent of the errors involved a secondary diagnostic process failure, for example: failure or delay in considering the diagnosis and putting too much weight on a competing diagnosis or failure or delay in ordering needed tests and even considering the diagnosis. When only looking at the 162 errors that were considered major diagnostic errors, 43% were related to clinician reasoning [3]. Generally it is not a lack of knowledge that leads to diagnostic error, rather it is the clinician's diagnostic reasoning process.

"The physicians of the future will not rely as much upon what they know as on their ability to find out about things they don't know.... Are we helping students to understand that rote knowledge is not the route to being a great doctor, it's knowing what you know, knowing what you don't know and knowing how to bridge that gap."

Steven Nissen

What are the reasoning processes clinicians employ? In their daily practice, the dual process model of clinical reasoning is routinely used. System 1 (or type I) rapid pattern recognition reasoning strategies are implemented for familiar problems. System 1 thinking is fast, unconscious, contextual and automatic. When the clinician is dealing with a complex or unfamiliar problem, system 2 (or type II) analytic thinking is utilized. It is deliberate and depends upon a systematic application of scientific understanding [4,5].

Comparisons of clinical reasoning processes of experienced physicians with those of novice clinicians, yields distinct differences. Pattern recognition or system 1 reasoning is strongly correlated with expertise. Presumably due to experience with multiple similar patients, expert clinicians "know" or "see" the diagnosis. Even though the development of expertise is dependent upon experience, studies in an array of fields have demonstrated that deliberate practice is required [6]. For the novice medical student or the resident whose clinical reasoning skills are continuing to evolve, to develop robust mental models of disease presentation and natural history requires thoughtful and guided practice. Skilled clinical faculty are needed to provide guidance and feedback on the medical student's or resident's progress. And, attention must be given to the types of patients learners work with to ensure that they have the opportunity to see, study and learn about a wide variety of diseases at various stages in their natural history, reflecting what they will encounter throughout their professional careers.

"Most of our decision-making is short, quick and thorough. We don't use all the information we just use a bit of the information. When you get so much information the way we then make it (decisions) is even more frugal, not the other way around. If you look at one's own life decisions, the biggest decisions are based on the least bit of information. The smallest decisions are based on the most amount of information, and people don't realize that. It's just the way we do it."

K. Ranga Rama Krishnan

Developing the mental models upon which both systems are predicated is necessary for the progression from novice to intermediate to expert and occasionally master level performance. However, even with robust mental models, errors occur. A recurring theme is that mistakes in clinical reasoning, seen with both type I and type II processing, are due to cognitive biases. While there are over 100 described biases, premature closure—terminating the assessment without obtaining critical information, availability bias—being influenced by a recent experience with a similar presentation but different diagnosis, and confirmation bias—seeking only information that will support the hypothesis, are among the biases commonly cited as leading to errors in clinical reasoning. As the magnitude and consequences of diagnostic and clinical reasoning errors have become increasingly recognized, a number of educational interventions seeking to counteract the presumed, albeit still controversial, pernicious effect of cognitive biases have been developed. Unfortunately, training physicians to recognize biases has shown mixed results but overall does not appear to reduce errors [7]. To date, while to a large extent self-evident, the best approach to minimizing diagnostic error is, as Daniel Kahneman stated,

- diligent clarification,
- thinking slow when confronted with a confusing picture, and
- humble self-reflection.

An interesting finding of a study based on real patient cases submitted to the Human Diagnosis Project showed that when groups of 2–9 randomly selected non-specialist physicians solved cases, their collective input resulted in increasing diagnostic accuracy. Individual specialist physicians solving cases in their specialty had a diagnostic accuracy of 63% whereas the accuracy increased to 86% for groups of nine physicians. Even when only two physicians participated the diagnostic accuracy increased to 78%. The implications of these findings remain to be fully explored [8].

As would be anticipated, given the heavy reliance on type I thinking by experts, the aphorism "we see what we know" is applicable in medicine. Studies have shown that primary care physicians entertained the correct diagnosis based solely on the chief complaint in 78% of patients [9]. Emergency room physicians developed 25% of their hypotheses before meeting the patient and 75% of their hypotheses in the first 5 minutes of the clinical encounter [10]. These findings highlight the positive value of heuristics or cognitive biases that allow efficient and accurate clinical reasoning.

The downside is that, as expertise is gained, there is a tendency to diagnose a disease or problem that resides within the physician's subspecialty. It is understandable that specialty physicians will be most apt to make a diagnosis within their subspecialty simply because the prior probabilities are high given the likelihood that the patient has been referred for consultation. However, the potential for misdiagnoses is raised if the patient has been inadvertently referred to the wrong subspecialty.

Years ago, I did think aloud study with physicians from various specialties. Individually, physicians were presented a case of an elderly man presenting with the acute onset of severe dyspnea with minimal exertion and exertional chest pain for 2 days. The physicians were asked what they were thinking after each bit of clinical data was revealed. Briefly, additional history was significant for vomiting up the prunes he had for breakfast the day his symptoms began and dental work 6 weeks earlier. The physical exam was remarkable for a 3/6 systolic murmur. The reasoning process of the various physician disciplines was interesting.

The cardiologists immediately diagnosed him with what was then called "crescendo angina" and that he needed to be admitted and have an emergent cardiac catheterization. Even when given additional data that his stools were 4+ guaiac positive and his hematocrit came back at 13%, their confidence in their diagnosis and recommendation did not change.

While the pulmonologists thought the most likely diagnosis was cardiac they cautioned that sometimes pulmonary problems can present in this manner and therefore a pulmonary etiology could not be discounted. When presented with the guaiac and hematocrit data, they quickly reordered their diagnostic possibilities, putting gastrointestinal bleeding as a high priority.

Although the gastroenterologists initially thought a cardiac etiology most likely, they cautioned that gastrointestinal bleeding could present with cardiac symptoms. Needless to say they were pleased with themselves when the blood in the stool and hematocrit data was revealed.

Perhaps the most interesting group were the general internists. They recognized a cardiac etiology as most likely but were the only ones to mention the antecedent dental work and concern that it might have resulted in endocarditis and a valve rupture. Likewise, they kept in mind the large prostatic nodule and that further assessment was in order after the acute situation was resolved. As additional data was presented, they reordered their diagnostic priorities. Seemingly they were juggling multiple diagnostic possibilities and bits of data that may or may not have been relevant.

As the actual patient's clinical course unfolded, it became apparent that what he thought were the breakfast prunes was actually hematemesis. With blood transfusions, all of his symptoms disappeared as did the murmur. He did not receive a cardiac catheterization.

Unfortunately, this problem of the "competency trap" is not limited to think aloud cases. It is encountered too frequently in clinical settings and likely is a major contributor to delayed or missed diagnoses.

"…knowing more about what you're competent in should take you to the risk of competency traps, which are traps where you turn everything into what you're competent in. What you don't spot is the instance where what you're competent in doesn't matter–it used to, it has in other cases, but it doesn't matter here. So if you think 'what I'm competent in takes care of every case' you're going to miss something important…. This is something that people figured out a long time ago. It's managing the tensions not resolving the tensions, that can't be resolved."

John King

Although correctly diagnosing a disease is vitally important, it is only one step in the clinical reasoning and decision-making process. When deciding which of several alternative therapeutic options to recommend, the clinician must take into consideration comorbidities, patient support systems, patient preferences and a host of "contextual" variables. Although characterized as the management plan, optimal clinical outcomes are as dependent upon the reasoning employed in its development as in the disease diagnosis. As the management plan is enacted and the disease course unfolds, if the patient's response is not consistent with expectations, the clinician must then modify the management plan and if indicated, reassess the accuracy of their initial diagnosis.

"It's the beginning of the (physicians) work to come up with the intervention and plan; and the plan is not a set of prescriptions and a return visit, the plan is about the patient accepting and participating in that process. We have to define our job that way because when the physician gets done writing the assessment and plan and handing it (to the patient) and thinking that their task is done, we know as a society the uptake isn't there."

Vikas Kheterpahl

Considered holistically, clinical decision-making is making sense of a series of data points that allows the clinician to attach a disease name. Then, as additional data points are considered, it is developing an approach to treatment and as treatment progresses making sense of the observed response. Clinical reasoning underpins what the physician does throughout the disease course until resolution.

As we ponder the implications for diagnostic clinical reasoning of developments in bioscience and technology, there are several considerations. As already discussed, more precise disease classification will become possible. However a cautionary note must be considered. Simply having more data points does not necessarily translate into more precise diagnosis. The "epidemic" of incidentalomas associated with the widespread use of so-phisticated imaging technology is a case in point. We know that individuals can harbor pathogenic staphylococcal aureus and not have clinical disease. We know that there is variable penetrance associated with genetic disorders. We know that individuals can have asymptomatic cardiac arrhythmias, such as premature ventricular contractions, that have no apparent clinical significance. How then do we educate our medical students and residents to know when to pay attention to a data point and when to ignore it? How do we avoid subjecting our patients to the emotional and physical suffering, let alone financial cost, associated with pursuing inconsequential "abnormalities"?

Currently clinicians know that the statistical definition of normal means that 1 in 20 will have an "abnormal" laboratory value. Intuitively the clinician assesses whether the abnormal laboratory value is consistent with the individual's overall picture and should be pursued, is of sufficient concern to repeat in a timely fashion or ignore. As the potential to intervene to restore to health before disease is manifest becomes a reality, how will we make the determination as to what requires intervention? Of the dozens or hundreds of potentially disease associated or even lethal gene mutations each one of us has, which should lead to action? How do we even decide what is normal? The problem will not be insufficient data points; it will be making sense of an overwhelming number of data points. While technology will provide sophisticated artificial intelligence and machine learning algorithms, it will not be a panacea.

Going forward, renewed emphasis should be placed upon diagnostic reasoning and the development of clinical judgment and decision-making skills [11]. It is crucial to teach students how to rationally approach diagnosis based upon an understanding of how a patient's risk factors and existing disease states affect prior probabilities; how the history and directed physical examination are utilized to increase or decrease the probability of a disease; and how to rationally select imaging, laboratory, and other diagnostic studies. One might think that this is what has been done for decades. Is there anything different? Our traditional approach to risk factors has been based on epidemiologic findings. As our understanding of health and disease has evolved, as represented in the model in Chapter 13, physicians are positioned to be more nuanced in their determinations of an individual's prior probabilities. However, simply having more data points will not necessarily lead to enhanced clinical diagnosis and decision-making.

What options are available to medical educators? One of the most powerful is having the student learner work with and care for actual patients, such as breast cancer patients, where the most sophisticated of diagnostic and disease progression assessments are part of the evaluation and decision-making. To provide a broader base of experience, the student or resident should work through instructive patient cases that incorporate genetic, microbiome, environmental and behavioral factors. They need to observe expert clinicians in practice and then hear how they think about a patient's presentation and disease course such as at a tumor board meeting or during teaching rounds. And importantly, as the novice medical student progresses in their education, the complexity of the clinical reasoning challenges should also increase. A beginning medical student will gain far more from working with and discussing a common presentation of a common disease whereas the senior resident will learn much more from the discussion of an esoteric case. A surgical resident will learn from a discussion of how a senior surgeon made sense of unanticipated complications encountered during a procedure whereas the medical student might not find it of value [12]. This requires the development of mental models with the breadth and depth necessary to set a robust and sophisticated foundation for clinical reasoning.

As Doug Paauw responded to the question "How do you train a diagnostician?"

"I think some of what we've always done is really help train diagnosticians. …You cannot just go with what's seen. What I mean by that is when you were in medical school and you were doing your medicine rotation, you would admit a patient and you would be grilled on the differential diagnosis, even if it was an 80-year-old patient with a COPD exacerbation, and it's clear as day it was a COPD exacerbation. You couldn't get away just saying, 'This is COPD exacerbation,' the attending would say, 'Well Jim, what else could it be?' 'Well um, um, well I guess this could be heart failure.' Okay, why is it or isn't it heart failure? And then, 'Well what else could it be?'

Now in our efficiency world, if you clearly know it's COPD exacerbation, there is no point to waste any of our time talking about what it isn't. Yet that's how you train a diagnostician because you realize that you're never going to recognize the atypical heart failure presentation unless you've thought about it. And then if you only think about it when it's typical, you only recognize the typical, and diagnosticians are able to see the rare manifestation of the common disease, and the common manifestation of the rare disease and the really good ones can recognize the rare manifestation of a rare disease. We're not even going to get the rare manifestation of a common disease if we don't push people to think of the possibilities."

Interestingly, explicit attention to developing medical students' [13] and residents' clinical judgment and decision-making remains less emphasized in most curricula as compared to knowledge acquisition. Given the explosion of biomedical knowledge and the data and capabilities that technologic developments will provide, explicit attention to developing the future physicians' clinical reasoning abilities is important.

Potential learning enhancements

- Have learners participate in the Human Diagnosis Project case challenges (www.humandx.org).
- As a patient's case is being discussed change key characteristics such as age, gender, past history and challenge the learners to discuss how this would alter their diagnostic thinking.
- In small groups access the biomedical literature and compile lists of known mechanisms resulting in common "diagnoses" such as diabetes, breast cancer, and congestive heart failure. Discuss how therapeutic approaches will change depending upon the underlying mechanism responsible despite sharing the same diagnostic label.

References

[1] American Diabetes Association. Diagnosis and classification of diabetes mellitus. Diabetes Care 2014;37(Suppl 1):S81–90. https://doi.org/10.2337/dc14-S081.

[2] National Academies of Sciences, Engineering, and Medicine. Improving diagnosis in healthcare. Washington, DC: The National Academies Press; 2015.

[3] Schiff GD, Hasan O, Kim S, Abrams R, Cosby K, Lambert BL, et al. Diagnostic error in medicine, analysis of 583 physician reported errors. Arch Intern Med 2009;169(20):1881–7.

[4] Norman G, Young M, Brooks L. Non-analytical models of clinical reasoning: the role of experience. Med Educ 2007;41:1140–5.

[5] Kahneman D. Thinking, fast and slow. New York, NY: Farrar, Straus and Giroux; 2011.

[6] Ericsson KA. Acquisition and maintenance of medical expertise: a perspective from the expert-performance approach with deliberate practice. Acad Med 2015;90(11):1471–86.

[7] Norman GR, Monteiro SD, Sherbino J, Ilgen JS, Schmidt HG, Mamede S. The causes of errors in clinical reasoning: cognitive biases, knowledge deficits, and dual process thinking. Acad Med 2017;92(1):23–30.

[8] Barnett ML, Boddupalli D, Nundy S, Bates DW. Comparative accuracy of diagnosis by collective intelligence of multiple physicians versus individual physicians. JAMA Netw Open 2019;2(3):e190096. [Accessed 2 March 2019].

[9] Gruppen LD, Woolliscroft JO, Wolf FM. The contribution of different components of the clinical encounter in generating and in eliminating diagnostic hypotheses. Res Med Educ 1988;27:242–7.

[10] Pelaccia T, Tardif J, Triby E, Ammirati C, Bertrand C, Dory V, et al. How and when do expert emergency physicians generate and evaluate diagnostic hypotheses? A qualitative study using head-mounted video cued-recall interviews. Ann Emerg Med 2014;64:575–85.

[11] Eva KW. What every teacher needs to know about clinical reasoning. Med Educ 2004;39:98–106.

[12] Cristancho S, Lingard L, Forbes T, Ott M, Novick R. Putting the puzzle together: the role of 'problem definition' in complex clinical judgement. Med Educ 2017;51:207–14.

[13] Rencic J, Trowbridge Jr. L, Fagan M, Szauter K, Durning S. Clinical reasoning education at US medical schools: results from a national survey of internal medicine clerkship directors. J Gen Intern Med 2017;32(11):1242–6.

Additional resources

- National Academies of Sciences, Engineering, and Medicine. Improving diagnosis in health care. Washington, DC: The National Academies Press; 2015. https://doi.org/10.17226/21794.
- National Academies of Sciences, Engineering, and Medicine. An evidence framework for genetic testing. Washington, DC: The National Academies Press; 2017. https://doi.org/10.17226/24632.
- National Academies of Sciences, Engineering, and Medicine. Biomarker tests for molecularly targeted therapies: key to unlocking precision medicine. Washington, DC: The National Academies Press; 2016. https://doi.org/10.17226/21860.
- Tcheng JE, Bakken S, Bates DW, Bonner H III, Gandhi TK, Josephs M, et al., editors. Optimizing strategies for clinical decision support: summary of a meeting series. Washington, DC: National Academy of Medicine; 2017.
- Human Diagnosis Project: an online consulting and education program to enhance medical training and decision making featuring patient cases inputted from around the world. https://www.humandx.org.
- Choosing Wisely. Promoting conversations between patients and physicians. http://www.choosingWisely.org/.
- Kerr EA, Kullgren JT, Saini SD. Choosing Wisely: how to fulfill the promise in the next 5 years. Health Aff (Millwood) 2017;36(11):2012–8.
- Kahneman D. Thinking, fast and slow. New York, NY: Farrar, Straus and Giroux; 2011.
- Groopman J. How doctors think. New York, NY: Houghton Mifflin Company; 2007.
- Society to Improve Diagnosis in Medicine. https://www.improvediagnosis.org.
- Lehrer J. How we decide. Boston, MA: Houghton Mifflin Harcourt; 2009.

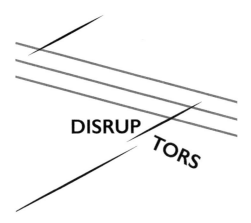

S E C T I O N III

We have considered aspects of physicianhood that are enduring and must remain central to the education of physicians regardless of breakthroughs in technologic advances and in our understanding of health and disease. Now let us turn our attention to potential disruptors. As defined in the Cambridge English dictionary a disruptor is: "a person or thing that prevents something, especially a system, process, or event, from continuing as usual or as expected" [1]. Based on this definition, identifying a development as a disruptor carries the assertion that it will greatly affect or fundamentally change the practice of clinical medicine.

Moving forward, I recognize the wisdom in the aphorism famously attributed to Yogi Berra, "It's tough to make predictions, especially about the future." History is replete with failed predictions from eminent scientists, acclaimed industrial leaders and authoritative sources.

To name but a few:

"Fooling around with alternating current is just a waste of time. Nobody will use it, ever."
Thomas Edison (1889)

"These so-called X–rays are only a deliberate hoax."
William Thompson, Lord Kelvin (1895)

"Who the hell wants to hear actors talk?"
H.M. Warner, Warner Brothers (1927)

"A rocket will never be able to leave the Earth's atmosphere."
New York Times (1936)

"I think there is a world market for maybe five computers."
Thomas Watson, chairman of IBM (1943)

"Television won't last because people will soon get tired of staring at a plywood box every night."
Darryl Zanuck, movie producer, 20th Century Fox (1946)

"If excessive smoking actually plays a role in the production of lung cancer, it seems to be a minor one."
W.C. Hueper, National Cancer Institute (1954)

"The world potential market for copying machines is 5000 at most."
IBM, to the eventual founders of Xerox, saying the photocopier had no market large enough to justify production (1959)

"We don't like their sound, and guitar music is on the way out."
Decca Recording Company on declining to sign the Beatles (1962)

"There is no reason for any individual to have a computer in his home."
Ken Olson, president, chairman and founder of Digital Equipment Corporation, (in a talk given to a 1977 World Future Society meeting in Boston)

"Two years from now, spam will be solved."
Bill Gates (2004)

"There's no chance that the iPhone is going to get any significant market share."
Steve Ballmer, CEO Microsoft (2007)

Recognizing that acclaimed scientists, business leaders and futurists (in addition to weather forecasters and financial pundits) are often wrong, what is the evidence, beyond speculation, that the practice of clinical medicine as we now know it will actually be disrupted rather than simply evolve? Is there anything different now? Is there any reason to significantly change our proven medical education approaches? Or, in 30 years, will we chuckle at the hubris to even think that the practice of clinical medicine will be disrupted?

In the chapters that follow, we will consider trends that I predict will affect the practice of clinical medicine and increase the likelihood of a disruption akin to what has forever changed the retail business, banking and financial transactions, and the overall dissemination of information.

Reference

[1] https://dictionary.cambridge.org/us/dictionary/english/disruptor. (Accessed March 13, 2019).

7

Demographics

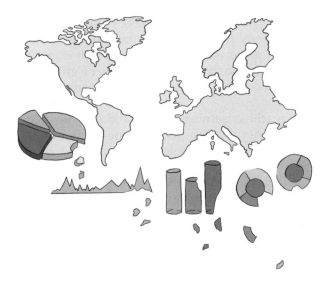

The demographics of the United States and of the world are changing. Of all the assumptions regarding what will affect the practice of clinical medicine mid-century, this is the most incontrovertible. Some demographic trends will require medical educators to modify and expand existing curricular offerings. Others may lead to changes that require physicians to develop new skills and approaches to providing care. Let us consider several examples.

Aging populace

The population in the United States and in other developed countries will continue aging. It is projected that by 2030 there will be more people over age 60 than under age 15 in North America. The numbers of

Implementing Biomedical Innovations into Health, Education, and Practice
https://doi.org/10.1016/B978-0-12-819620-5.00007-2

elderly individuals will continue to rise and those age 80 and above will grow disproportionately, accounting for nearly 10% of the total population in the developed world by 2050 (estimated at 379 million) [1]. This has many implications for physicians as well as for society, including an increase in the total amount of disability, ill health and frailty that will require care.

The economic implications of an aged population are well known, with clinical and assistive care comprising a significant portion of that cost. While recently questioned, a recurring observation is that the most medical resource-intensive period is the last 6–12 months of life [2]. This is not limited only to the aged. Discussions regarding how best to deploy finite medical resources raise questions about the appropriateness of expensive, but ultimately futile, medical interventions at the end of life. While for study purposes, once a person has died, it is easy to determine the resources expended on their medical care over a given period of time, the difficulty is assessing prospectively when an individual is in the last stages of their life.

Advances in machine learning will allow ever more accurate estimations of life expectancy. By incorporating information on comorbidities in patients with cancer, machine learning-developed algorithms have already demonstrated the potential to enhance the accuracy of survival predictions [3]. It is expected that these algorithms will also improve the accuracy of survival predictions for patients with multiple chronic diseases and trauma. From a positive perspective, this will allow physicians to have realistic, albeit difficult, conversations with patients and their families regarding care options and preferences. Providing a more accurate estimate as to how long an individual has to live will inform decisions regarding hospitalization, procedures, and all medical choices as well as their personal lives and priorities.

The aging of the populace has multiple potentially negative ramifications. As the ratio of the aged, dependent population to the adult working population increases, the strain on governmental and social resources could result in societal schisms. It may lead payers, whether governmental or private, to expect physicians to limit care in the last stages of life, raising a multitude of ethical issues. If the institutions through which the physicians are paid are reimbursed on a per capita or a shared savings model, additional pressures might be exerted to limit diagnostic and therapeutic expenses, raising additional concerns regarding professionalism. The economic burden of providing care to an aged population will force physicians to explicitly consider cost and life expectancy when recommending a course of clinical action. A well-developed grounding in ethics and the principles that form the basis of our profession will be necessary.

"Aging in place is really taking a very holistic look at the environment and the intersection of environment and humans to say 'How do we do this?' The biggest growth in real estate now is not in the retirement communities anymore, it's redesigning homes to be able to accommodate what's needed for aging. The groups that I think are going to have very central roles here are therapists, occupational therapists, at least as I've known them, and physical therapists."

Mary Naylor

Aging and the attendant disability and cognitive decline often leads to institutionalization in a nursing home or other facility for the elderly or requires family intervention and care. Not only is institutionalization financially costly, aged individuals frequently express the wish not to be institutionalized, but rather to spend their time in familiar settings with familiar people. This has led to the concept of "aging in place" which is increasingly enabled through artificial intelligence linked sensors and robots. Entrepreneurs are responding with an array of inventions designed to maintain and enhance function and support independence [4].

Coupled with virtual communication capabilities, a suite of technologic innovations will allow individuals to continue living independently, will provide family members and caregivers the ability to monitor and ensure the safety and well-being of the individual, and will decrease the need for "institutional beds." While enabling aging in place with technologic assistive devices is desirable from the individual's perspective, as discussed in the section on ethics, these technologic devices can also be intrusive and they raise important ethical questions.

Regardless of the options provided by innovative technology, the need for clinicians skilled in the care of the aged will increase. As longevity increases, unless age-related physical and cognitive decline is delayed, caring for this cohort will be the focus of much of the clinical workforce. This includes physicians as well as a wide range of caregivers skilled in managing aging in place and dedicated to maximizing function in the elderly.

Due to economics, personal preference, and increased risks of complications in hospitals, individuals will often opt for treatment in their homes rather than in clinical care facilities.

Across the age and social economic spectrum, there is a growing trend to provide "hospital at home" (HaH) care. Increased public awareness of contagion, financial considerations, and the desire to spend days at home rather than in a hospital, are fueling this trend. Infectious disease treatment, infusion therapy, dialysis, and rehabilitation following surgery and trauma are but a few examples of therapeutic treatments that will preferentially occur in the individual's home. In a retrospective analysis of patients receiving HaH care who subsequently died within 6 months, when compared to a matched cohort, the HaH patients spent an average of 20 more days at home [5]. Pilot studies have demonstrated that adults with infection, heart failure exacerbation, chronic obstructive pulmonary disease and asthma, randomized to HaH or standard inpatient hospital care, showed a 67% lower median direct cost for the acute-care plus 30-day post discharge period for HaH treated patients as compared to standard in-hospital care. HaH patients were nearly three times more physically active during the acute phase and had fewer readmissions (11% versus 36%). The outcomes of this pilot study showed no significant differences in quality, safety, patient outcomes or patient and family experience [6]. Similarly, HaH bundled with 30 days of post-acute transitional care resulted in improved patient outcomes and better ratings of care [7].

While HaH care is currently heavily dependent upon in person professional care, the innovative technology that enables aging in place, augmented by monitoring equipment and robotic devices, makes this option feasible for an ever-growing array of conditions, ranging from acute disease intervention to rehabilitation. In addition, this will require development of the technology and technologic support workforce as well as of the clinical professionals skilled in facilitating patient care in home settings.

Obviously, this has negative financial implications for the current business model of many hospitals.

Geographic upheaval

Geopolitical disruption, immigration and international travel will increase the likelihood of disease spread globally.

Armed conflicts, natural and manmade disasters that displace people, international travel and the movement of goods will continue the trend toward rapid global transmission of disease. Whether it be well known diseases such as influenza, historically regionally contained diseases such as Ebola or previously unknown diseases such as SARS or Zika virus, infectious diseases know no geopolitical boundary and clinicians must be prepared to consider all possibilities. Similarly, due to the ease of global travel, individuals coming to the United States from around the world may present with signs and symptoms of diseases from their region of origin that are "foreign" to US-trained physicians. Schistosomiasis, leprosy, and rheumatic valvular disease are examples of maladies that occur commonly in some regions of the world but may not be part of most US medical students' and residents' clinical experiences.

This has important implications not only for disease surveillance, but also for implementation of machine learning algorithms. As discussed in depth in Chapter 8, as machine learning algorithms are employed to aid in diagnosis, it is critical that the learning database include individuals from the population from which the patient hails. In addition, while a travel and exposure history is routinely taught as part of clinical skills and diagnosis curricula, it frequently is overlooked in clinical settings. Medical educators and clinical mentors need to reinforce how prior probabilities for diagnosis may be altered by information regarding travel and exposures.

Diversity

The increasing ethnic and cultural plurality of the populace has direct implications for clinical care and thus for medical education. While the importance of "cultural competence" has been understood for decades, our ability as medical educators to provide students and residents a framework for optimizing work with patients and families from widely differing backgrounds remains a work in progress. We must instill in our students an understanding of the importance of respecting, and not violating, norms based on ethnic, religious and cultural perspectives for a wide range of clinical situations ranging from every day communication styles, to words used to describe and ascribe signs and symptoms, to whom to involve regarding clinical decisions, to the approach to discussions regarding end-of-life treatment. All too frequently the unstated assumption has been that everyone has the same expectations for the role of physicians, appropriate communication patterns, concepts about health and disease, and the ethical framework that should guide decision-making. This is not true. The multiplicity of ethnic and cultural backgrounds of our patients means that differing worldviews and expectations will be encountered by our students and residents in their professional practices. Future physicians and thus also their medical educators will need to understand and acknowledge racial, ethnic and cultural differences when dealing with subjects ranging from communication skills to diagnosis and therapeutic options.

Economic divide

We are also becoming increasingly more economically divided. The implications of the economic divide for clinical medicine are multifold. Those with the financial means might expect or even demand services, procedures, and new technologic interventions. The historical ethos, whether or not thought to be true, that medical care would be based only on clinical indications and not on financial considerations will be stressed and will further confound the complexity of ethical decisions.

Differences in social economic status also have major implications for the development of disease and erosion of health. Factors important to health including diet, housing, educational opportunities, and even access to clinical professionals are imperiled by lower financial resources.

Imbuing our students and residents with the responsibility to care for all, regardless of social or economic status, will remain an essential role for medical educators. Further, it means that our academic medical centers must adhere to this philosophy and provide care to all regardless of financial status. Physicians cannot become 'service providers' only to those with financial means.

Urban/rural

The continued growth of urban populations has many implications for health and disease. The built environment in many urban centers has not been constructed with health in mind. Zoonotic infections, sanitation concerns, safety, crowding, noise and light pollution are but a few of the potential adverse consequences of many urban environments. Often this is exacerbated as new migrants to the urban centers live in crowded, dilapidated housing and have inadequate access to healthy food options and medical care. Homelessness in our urban areas and displacement due to gentrification of neighborhoods present significant challenges to health and well-being.

The rural population is increasingly underserved with inadequate access to medical care. Technologic constraints due to low population density means that Internet access is limited and the opportunity for the rural population to participate in technology-enabled care, including online sensors, is limited. Moreover, the plight of the rural poor is often times less visible as compared to the urban poor and they have been disproportionately affected by some modern plagues such as methamphetamine and opioid addiction.

While some medical schools have consciously established a curriculum to educate students in the unique medical needs and care environments that rural communities present, the need is greater than the supply. Moreover the professional opportunities, economic conditions, cultural and lifestyle realities in rural areas of the nation are not viewed as desirable by many young physicians.

Just as economics divide, so too does the urban/rural lifestyle. As technology enables asynchronous care, it holds the promise of addressing the needs of rural as well as inner-city urban populations. Whether or not this promise is met will depend upon educating the workforce that can effectively utilize the technology, a task for medical educators, coupled with policy and regulatory decisions that enable its widespread deployment.

To state the obvious, as medicine exists to address human needs in time of sickness and disability, demographic trends influence clinical practice. Medical educators must be cognizant of these trends as they will require modifications of existing curricula. As noted by Clayton Christensen, "The breakthrough innovations come when the tension is greatest and the resources are most limited. That's when people are actually a lot more open to rethinking the fundamental way they do business." The confluence of demographic trends with technologic innovations will increase the likelihood of widespread adoption of the technology as it addresses otherwise unmet needs. The changes predicted based on demographic trends will require physicians to develop new skills and medical educators to prospectively modify curricula to meet the needs of the populace.

References

[1] Harper S. Economic and social implications of aging societies. Science 2014;346(6209): 587–91.

[2] Einav L, Finkelstein A, Mullainathan S, Obermeyer Z. Predictive modeling of US health-care spending in late life. Science 2018;360:1462–5.

[3] Zolbanin HM, Delen D, Zadeh AH. Predicting overall survivability in comorbidity of cancers: a data mining approach. Decis Support Syst 2015;74:150–61.

[4] Kohlbacher F, Herstatt C, Levsen N. Golden opportunities for silver innovation: how demographic changes give rise to entrepreneurial opportunities to meet the needs of older people. Technovation 2015;39–40:73–82.

[5] Zimbroff RM, Leff B, Siu AL. Hospital at home–plus reduces days spent in hospitals and other inpatient facilities. NEJM Catal A 14, 2018. Downloaded May 19, 2018, https://catalyst.nejm.org/hah-plus-days-spent-hospitals-home/.

[6] Levine DM, Ouchi K, Blanchfield B, Diamond K, Licurse A, Pu CT, et al. Hospital level care at home for acutely ill adults: a pilot randomized controlled trial. J Gen Intern Med 2018;33(5):729–36. https://doi.org/10.1007/s11606-018-1307-z.

[7] Federman AD, Soones T, DeCherrie LV, Leff B, Siu AL. Association of a bundle hospital-at-home and 30-day postacute transitional care program with clinical outcomes and patient experiences. JAMA Intern Med 2018;178(8):1033–40.

Technology and computing

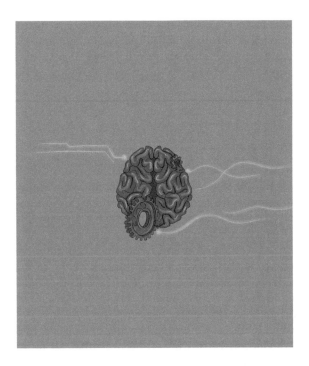

We know that technologic and computational advances over the last half-century have touched virtually every aspect of society and modern life. The disruptions that have resulted in areas as disparate as communications, banking, fraud detection, retail sales and marketing are well known. Clinical applications ranging from monitors to algorithmic interpretation of EKGs to electronic health records have proliferated. However, within the last 10+ years, advances in computational power, miniaturization and technologic innovation have resulted in a dramatic increase in the numbers and potential impact of clinical applications. Will the practice of clinical medicine be as disrupted as other industries?

Implementing Biomedical Innovations into Health, Education, and Practice
https://doi.org/10.1016/B978-0-12-819620-5.00008-4

"If one looks ahead 20 to 25 years from now, we are going to see dramatic change. If you consider what is rapidly advancing in data analytics, the power of computers, machine learning, AI, wearable devices that monitor our health status, healthcare will change from a reactive science of diagnosing and treating disease to one where we will be able to predict disease before it happens. What some of us call Predictive Health.

So the role of the physician will change immensely. For example, while society will still need trauma surgeons they will be assisted by robots, virtual reality, that will help them do it better. But I think when it comes to disease management it's really going to move way up the ladder to managing changes in our metabolic pathways before any disease becomes manifest. So we will need different kinds of doctors who will be monitoring heath status and will be counselors and empathetic healers, guiding people in new ways to manage changes in their 'omic' pathways that prevents disease from manifesting."

Michael Johns

Perspective

Technologic advances have played a central role in all phases of the journey from antiquity to present-day medicine. Everything from fundamental principles of health and disease to medical disciplines to the structure of medical education has been shaped by technology.

Advances in our understanding of the basis of health and disease have been in large part predicated on advances in technology. The invention of the microscope by Anton van Leeuwenhoek in the late 17th century led to the discovery of bacteria. Beginning in the mid-1800s, advances in microscopy led to the discovery that a wide range of diseases were caused by bacteria and the development of the germ theory. Similarly, microscopy advances resulted in increased understanding of diseased and healthy tissues and the development of the cell theory—that all living organisms were composed of units invisible to the naked eye termed cells. Equipment and techniques from the nascent field of chemistry were applied to biological problems in the late 1800s, resulting in revolutionary advances in our understanding of physiology and perturbations thereof in disease states.

While Gregor Mendel is thought of as the father of genetics for his work on pea plants in the mid-1800s, it was not until the early 1900s that his work was rediscovered and replicated and the scientific study of genetics was embraced. Although the tendency of some diseases to occur in kindreds was recognized and the mode of inheritance established, the mid-20th century discovery of the role of deoxyribonucleic acid (DNA) in genetics by James Watson and Francis Crick was required before the specifics of genetic defects could be identified. The technique of polymerase chain reaction (PCR) developed in the early 1980s and the rapid advances in automation predicated on this technology led to the identification of thousands of genetic defects that result in disease or increase the likelihood of disease. New insights into the role of genetics in non-communicable diseases, ranging from rare to common [1], hold the potential for dramatic changes similar to that seen with our knowledge of the role of microbes in infectious diseases.

Technologic advances not only led to breakthroughs in our scientific understanding, they also resulted in incredible improvements in diagnosis and therapy. One example, the production and detection of the X-ray by Wilhelm Roentgen in the late 1890s revolutionized clinicians' diagnostic capabilities. It also resulted in the development of the specialty of radiology. Subsequent technologic breakthroughs have led to a number of imaging techniques ranging from plain X-rays to magnetic resonance imaging (MRI) to ultrasonography to positron emission tomography (PET) scanning. In turn, this has allowed clinicians to diagnose a multitude of diseases with ever greater precision and rapidity.

Physicians of my vintage likely remember that a stop in the radiology department to examine films of our hospitalized patients, discuss their findings with the radiologist and provide feedback on patients previously reviewed was an obligate part of rounds. The reading room was a center of clinical activity and patient discussion. Technology has dramatically altered this scenario. The introduction of picture archiving and communication systems (PACS) not only replaced conventional radiological film, it meant that images were widely available and distributed across the network. Consequently, physicians no longer needed to go to the reading room to review their patients' imaging studies. An unintended consequence is that the dialog between the radiologists and the physicians caring for patients rarely occurs. This decreases the opportunities for radiologists to make follow-up recommendations and has eliminated the feedback loop to the radiologists. While this is clearly a technologic advancement from the perspective of image handling and availability, the negative consequences on learning, on the part of both the radiologist's and the clinicians directly caring for the patient, are significant and unanticipated.

Early in the 21st century, we are beginning to appreciate the fundamental changes to the practice of clinical medicine that recent technologic and scientific advances envisage. As we consider the potential impact of various innovations, it behooves us to consider the role they are playing (Table 8.1). Many technologies allow us to precisely quantify biometrics, better measure known physiologic variables such as blood pressure, cardiac rhythm, and cells and proteins in blood. Others, building on known physiologic processes, provide interventions such as cardiac pacing. Still others are the tools of scientific discovery, the applications of which to

TABLE 8.1   Technologic innovations that will alter the practice of clinical medicine in mid-21st century

All clinical studies that generate digital data will be automatically "read and interpreted." This ranges from imaging studies to pathologic specimens and physiologic tracings.

During surgery and all procedures involving sedation, physiologic monitoring, anesthetic administration, and interventions to maintain appropriate temperature, blood pressure, perfusion, etc. will be automated.

Once a patient is intubated, respirator settings such as percent 02, pressures, weaning parameters and when extubation can be safely accomplished will be automated.

Open surgery will essentially disappear. Endoscopic approaches will be used for pulmonary, gastrointestinal and genitourinary diseases. Solid organ and internal problems will be addressed laparoscopically or via focused ultrasound to ablate tissue such as tumors. Robotic surgical instruments coupled with real-time imaging will replace human operators for laparoscopic and endoscopic surgeries.

Gene editing techniques will be used to "treat" and prevent diseases caused by genetic disorders.

Pharmacologic therapy will be tailored to an individual's unique genome and microbiota ecosystem as modified by their environment and behavior.

Circulating free tumor DNA and malignant cells will be used to diagnose and determine tumor sensitivity to pharmacologic agents. This will allow monitoring of tumor evolution and guide therapeutic modifications.

Point-of-care technology will enable rapid diagnosis of sepsis and identification of appropriate therapeutic agents.

Sensors will be utilized to monitor individuals with chronic conditions facilitating individualized therapeutic regimens and interventions early on when control is compromised.

health and disease remain to be determined. And some enhance human capabilities ranging from communication to information processing to recurring physical tasks.

"… The whole future of technology and the reason I got so excited about it is that I see it as a humanizing thing that will relieve the human drudgery, the machine will do what it does best and that is enrich the opportunity for the human to do what they can do best."

Daniel Atkins

Table 8.1 lists only but a few of the changes that technology will likely enable in clinical medicine. But not all technologic advances prove beneficial. As we consider broad categories of impact, it is incumbent upon us to approach each with the understanding that medicine is littered with stories of "advances" that have proven to be harmful or have not lived up to their initial promise. With the advantage of hindsight, we are bemused that esteemed physicians would develop machines to induce gastric hypothermia, or as it was termed "gastric freezing," as a treatment for duodenal ulcers [2]. However, thousands of patients were treated with attendant morbidity and mortality before it was discredited. Regrettably patient morbidity and mortality associated with treatment advances is not limited to distant historical examples. The recent experience with power morcellation, used primarily during laparoscopic hysterectomy or myomectomy procedures, is a case in point. In women with unanticipated uterine cancer, the use of morcellation during the surgical procedure risked dissemination of cancerous cells leading to worsened prognosis [3]. These are examples of invasive technologies where the untoward outcome is readily apparent and clearly linked to the procedure. As we consider future advances it is likely that adverse outcomes will not be as readily identified nor linked. Ever more vigilance and caution will be needed as the benefits and risks of technologic "advances" are weighed.

Several of the educational recommendations set forth in this book emanate from advances in technology that are predicted to have significant effects on clinical practice. But what is the evidence that these technologic innovations will occur? Does there appear to be momentum that justifies the prediction, let alone require substantial changes in our pre-medical and medical education systems? Or, is it simply conjecture that may sometime in the distant future come to pass but is unlikely to have much of an influence on mid-century clinical medicine?

Let us look at the evidence in a few areas. The examples and the cautions noted are not comprehensive but exemplary.

Advances in computing power and miniaturization have enabled a wide array of technologic innovations. Some are primarily sensors such as in fitness trackers. Others are primarily computational, such as search engines. Still others are sensors embedded in mechanical devices linked by computer programs such as the da Vinci surgical robot. While I will consider each area separately, it is an artificial construct as there is a continuum of integration.

Sensors

"The sensor thing is interesting, the phone can do so much already, but there's going to be additional sensors and probably your home is going to be impregnated with them, your car will be impregnated with them. When you sit in your car you could be having a clinic visit because of the amount of information that is taken up - your heart rate and blood pressure, all the physiological parameters.

Your voice will be able to tell your sentiment and your stress levels. It will almost be like a hemoglobin A1C for your emotional health, for mood. So there's all of these very exciting things and I think that the obligation of the doctors of tomorrow is to contextualize that for a patient."

Sanjay Gupta

Dramatic technologic advances have made an array of sensors not only a potential but a reality. Consumer products that provide activity and basic physiologic measurements—such as fitness trackers, smart watches, and smartphone apps—have burgeoned in popularity. Wearables that can measure activity, caloric expenditure, temperature, oxygenation, blood pressure, heart rate and electrocardiogram (ECG), electroencephalogram (EEG), electromyogram (EMG), sleep patterns, balance and gait are available.

To date, the major focus of consumer oriented wearables is to enhance healthy behaviors. Devices that measure and improve health behavior are estimated to have been purchased by up to 20% of Americans, but research is showing that approximately 1/3 of those discontinue using them within 6 months of purchase. Moreover, the efficacy of wearables to facilitate healthy behavior change such as increased activity is spotty at best. Not surprisingly, personality factors appear to be as or more important than the wearable technology in predicting increased exercise levels.

A natural progression has been to use this sensor technology for clinical purposes ranging from diagnosis to chronic disease monitoring. One application is to diagnose cardiac arrhythmias. A recently reported randomized clinical trial studied over 2500 participants at risk for atrial fibrillation. The intervention was wearing a self-applied continuous ECG monitoring patch while at home and during routine activities for up to 4 weeks. Atrial fibrillation, a common arrhythmia associated with an increased risk of stroke, was diagnosed in 3.9% of the monitored individuals as compared to 0.9% in the non-monitored control group. Of note, despite volunteering to participate in the study, approximately one third of the individuals randomized to wear the electrocardiogram patch did not do so [4].

Recently reported was a study of over 419,000 self-enrolled people who did not have a diagnosis of atrial fibrillation and were not taking anticoagulants to determine the accuracy of atrial fibrillation detection using Apple Watch photoplethysmography technology. If the watch embedded sensor detected 5/6 irregular pulse tachograms within a 48-h period, the individual was notified and asked to contact the investigator for a video interview and potentially wear an ECG patch for up to 7 days. Confirmation of atrial fibrillation was based on simultaneous recording of atrial fibrillation on the ECG patch and the watch-based tachogram. Two thousand one hundred sixty one participants received notifications. Six hundred fifty eight individuals were sent ECG patches of which four hundred fifty were returned for analysis. Thirty four percent of returned patches confirmed atrial fibrillation. The watch-based tachogram had a positive predictive value of 71%. Interestingly, while about half of the participants who received the irregular pulse notification contacted the study doctor, follow-up surveys showed that only 57% of those sought medical care with their personal physician outside of the study [5].

Although not as accurate as the "gold standard" ECG, this technology shows promise for the detection of atrial fibrillation and it is anticipated that the accuracy of the sensor will continue to be improved. But what does that mean for physicians? As individuals with previously undiagnosed or even undetected atrial fibrillation are identified will the presumed natural history and risk for atrial fibrillation require rethinking? Will it be possible to risk stratify people for stroke and target intervention to a subset or will all require anticoagulation or another therapeutic approach? Beyond the implications for physicians interacting with their patients, what are societal implications? Given our current understanding that atrial fibrillation increases in prevalence with age, should all citizens who reach a certain age be given the option or be required to wear an appropriate sensor?

The detection of intermittent atrial fibrillation in individuals who are otherwise asymptomatic is but one example. Our understanding of the implications of other arrhythmias is likely to be similarly challenged. It is likely that the accuracy and sophistication of wearables and smartwatch based sensors will allow the collection of data regarding cardiac rhythms from hundreds of thousands or millions of people. This will present a rich opportunity for research but will also require rethinking as to intervention thresholds and therapeutic recommendations. And cardiac rhythms are but one example of the multitude of physiologic parameters that sensors will be able to measure in real time. The challenge will be how best to develop the cognitive models that will incorporate the new knowledge and, very importantly, be able to adjudicate in a rational way for one's patients what will sometimes be conflicting recommendations of different professional organizations.

But diagnosis using sensor generated data will not always involve physicians or even clinical settings. An ongoing quest has been to empower individuals to accurately self-diagnose without professional assistance or intervention. One such initiative was the Tricorder competition [6]. Combining self-entered and sensor-derived data with computer-based algorithms, prototypes have been developed that allow accurate diagnosis of common conditions including "normal—no disease" without the involvement of clinical personnel. Such initiatives will only proliferate. While the benefits of avoiding unnecessary clinical or emergency room visits are readily apparent these devices will bring with them a wide array of unique and previously not addressed regulatory, calibration and validation, and updating issues. One can begin to imagine the accusations of restraint of trade and protectionism that will arise when physicians rightfully advocate for governmental and regulatory oversight to protect the public from substandard devices.

Beyond diagnosis, powerful applications will leverage the innovations in sensor technology to facilitate chronic disease monitoring, therapeutic regimen compliance and care delivery. Applications across an array of chronic diseases including chronic obstructive pulmonary disease, asthma, sleep apnea, congestive heart failure, cardiac arrhythmias, hypertension, obesity, rheumatoid arthritis, low back pain, Parkinson's disease, and bipolar depression are among those studied. Proponents emphasize the potential to improve patient outcomes, reduce utilization and costs and generate vast amounts of data for research. Anecdotally, biosensors have shown some efficacy. However, well-designed clinical trials while few in number, have shown mixed results. Noah and colleagues, in a meta-analysis of randomized controlled trials of wearable biosensors, found "early promise in improving outcomes for patients with select conditions, including obstructive pulmonary disease, Parkinson's disease, hypertension and low back pain. Interventions aimed at increasing physical activity and weight loss using various activity trackers showed mixed results" [7]. The most common study design employed a physician or other clinical professional to assess the sensor data and provide instruction and feedback to the patient. Small sample size and short study duration were limitations that decreased the likelihood of finding significant effects. Despite these mixed results, it is expected that for patients with selected chronic diseases, continuous monitoring with frequent therapeutic adjustments will replace the current practice of episodic monitoring and adjustment of therapeutic regimens.

While one would assume that individuals with chronic disease would welcome the addition of sensors to allow detection of changes requiring intervention, the lackluster experience with sensors as currently constituted in wearables and in the above mentioned atrial fibrillation diagnostic study should not be discounted. Sensors must be used to be useful. While refinements in the technology will continue apace, physicians and clinical caregivers will need to ensure that the patients who will most benefit from continuous monitoring do so.

While the passive collection of data on patients with an array of chronic diseases enables feedback to the patient and valuable information for clinical management and follow-up, linking sensors with delivery devices in a closed loop therapeutic delivery system is a logical next step. The development of such systems to manage blood glucose levels in patients with diabetes mellitus has been an ongoing focus for decades [8]. Common to all is a sensor, a decision-making algorithm and an insulin delivery mechanism such as an insulin pump. While seemingly a straightforward problem that should readily lend itself to a technologic solution, the impediments encountered when developing such a device are instructive. Firstly, a comfortable, non-obtrusive and reliable sensor that can be worn 24 h a day 7 days a week whether passively watching TV or swimming must be developed. Then, the algorithm that takes the signal from the sensor and translates it into commands to the pump must be written. This algorithm has to accommodate a wide range of physical activity, meals, sleep, and hormonal and psychological stressors. Given the real possibility of harm from hypoglycemia or ketoacidosis, the algorithm and its control system must be fault-tolerant and include detection and diagnostic modules for problems ranging from infusion pump blockage to sensor failure. Similar challenges will be encountered across the spectrum of diseases where such a closed loop therapeutic delivery system has the potential to enhance control and patient outcomes.

Increasingly, the data generated by sensors, broadly defined, is being used to identify biomarkers. When coupled with the power of machine learning, the massive amounts of information generated will potentially lead to discoveries of physiologic biomarkers predicting complications and disease progression and will point to interrelationships among diseases and physiologic changes. The magnitude of the data currently available through sensors is exemplified by a recent study that recorded over 250,000 daily measurements in 43 people. Seven sensors per person measured heart rate, peripheral capillary oxygen saturation, skin temperature, sleep, steps, walking, biking, running, calories expended, acceleration due to movement, weight, gamma and x-ray radiation exposure. Analyses of the data showed differences in individuals' physiologic circadian rhythms, detected changes due to incipient infection, effects of high-altitude airline travel, and significant differences among individuals in physiologic and activity profiles [9].

As the focus of innovation moves from treating evident disease to detecting a disease at its earliest stages with the goal of restoring health, an important component of that future state will be the monitoring of physiologic variables and biomarkers through sensors. Historically, clinicians have not done a good job motivating individuals to change behaviors, including those that have clear negative consequences. Will we be more successful in encouraging apparently healthy individuals to utilize sensors and then make appropriate modifications when no change in health status is apparent? Conversely, will we be confident that the data generated is interpreted accurately and does not lead to unnecessary testing and psychological distress?

"...the biggest nut to crack is behavioral modification of people in a sustained and durable engagement. Maybe we can do that with sensors. The issue is that there are a lot of people resisting, they're worried about their privacy and infringement on their life. I've also learned about this concept of deliberate ignorance, that people don't want to know. There was a recent paper, the title of it's called Cassandra's Regret, and it showed how when people were asked if you could predict the future, illnesses, when you're going to die, and on and on, almost 80% of people said, 'I don't want to know, I don't want to know any of that stuff' [10]. Well, we have to get over that in order for people to be willing to be 'sensored.' So part of the problem is not wouldn't this be a wonderful opportunity for doctors, I think it's more at the individual level which is would they do it."

Eric Topol

Like all things technologic, the broad application of sensors is not a panacea. While the sophistication of existing sensors is remarkable, engineering must be cognizant of potential inherent design weaknesses. One example is the very common MEMS accelerometer. Thousands of devices, ranging from smart phones to implantable medical devices to automobiles, employ these accelerometers. A promising clinical application is using this technology to monitor the gait of patients with Parkinson's disease and their response to therapy. At the core of the accelerometer is a mass suspended on springs. When the device, which contains the accelerometer, changes speed or direction, the mass moves, generating a signal. However, the mass can also respond to the resonance generated by an acoustic interference and can falsely sense device movement [11]. While technologic and software solutions to this problem are being developed, one can imagine the potential harm that would come from clinical decisions based on erroneous data generated by an unreliable MEMS accelerometer-based sensor. Similar vulnerabilities will likely be discovered in other sensors but only after deployment in thousands of devices.

Another growing threat is cybersecurity. To date, clinical device manufacturers have largely neglected data security. Going forward, this must be addressed seriously to ensure that data being transmitted from sensors is not hacked. The potential for harm is real and manufacturers are increasingly focusing on security measures. Like many things, vulnerability to hacking and cybersecurity issues are a direct result of decisions made in decades past.

"The Internet was originally conceived as a very open environment. It was to permit collaboration and interaction... It was never designed to be hardened or robust against hacking. So we've been sticking on Band-Aids, and incremental fixes. The challenge of security and robustness is going to continue. ... The medical field can leverage heavily off of the autonomous vehicle because that is, as you know, life in the hands of the technology."

Daniel Atkins

The vast amounts of data that sensors generate require sophisticated computer programs to monitor and alert. These programs, often labeled artificial intelligence, are considered below.

Artificial intelligence/Machine learning

Online shopping recommendations, choosing stocks and market timing, language translation, autonomous vehicles, virtual assistants, real time driving directions, the current and envisioned applications of artificial intelligence (AI) are ubiquitous and highly variable. Defined as the development of computer systems able to perform tasks such as visual perception, speech recognition, and decision-making, normally requiring human intelligence, AI applications are increasingly common in medicine. Let us consider some of the areas that fall under the umbrella of artificial intelligence.

Machine learning

AI is sometimes used synonymously with machine learning in reference to the powerful computer programs that underlie these applications. But what exactly is machine learning? Machine learning is the intersection of computer science and statistics, using mathematical models applied to a dataset in order to identify patterns. Under the banner of machine learning are dozens of mathematical approaches, each with strengths and weaknesses. Simplistically, it means finding the curve that best fits dozens to thousands of input variables in order to predict a given outcome or cluster related variables using a complex mathematical model that can "learn" without being explicitly programed. In contrast to "expert system" computer-based algorithms that are rules based, machine learning makes no a priori assumptions as to which variables to include and the relevance of the resulting algorithm is judged based on the accuracy of its output. While anthropomorphic names such as neural networks, deep learning, cognitive computing, etc. are used and the models are said to emulate the way humans think, all are very complex mathematical programs.

At the most basic level, machine learning programs can be broken into three types—supervised, unsupervised, and reinforcement learning [12]. Supervised learning accounts for the majority of clinical applications to date. A simple way to think of supervised learning is that it uses a database with clearly defined inputs and outputs and then develops an algorithm that accurately predicts an outcome based upon similar inputs. An example might be input from a digitized mammogram and the output being an interpretation of a right-sided medial lower quadrant breast density consistent with cancer. To construct such an algorithm, a well curated data set consisting of hundreds of thousands of digitized mammograms including images from different makes and models of machines to minimize bias due to slightly differing radiation doses and processing technologies, with findings covering the entire spectrum of expected breast pathology, are fed into the computer program. The program then processes the billions of pixels of data and develops the algorithm that best predicts a given outcome. Subsequently a similar, but independent, data set is presented to the computer and the accuracy of the algorithm is assessed for the outcomes of interest.

A second group of machine learning programs employ what is called unsupervised learning. Massive heterogeneous data sets, without defined outputs, are entered and the computer program generates groupings or clusters of items culled from thousands of variables that share a relationship as determined by the computer developed algorithm. This approach is used to find structure among widely disparate variables and is useful for hypothesis generation. Take for example a seemingly heterogeneous condition such as autism spectrum disorder. A wide array of variables ranging from behavioral to genetic to physiologic are entered and the resulting heterogeneous database searched for patient clusters. Much of the work in precision medicine and learning health system initiatives is built upon the power that unsupervised machine learning algorithms provide.

The third type of machine learning programs, reinforcement learning, makes decisions sequentially, using trial and error, to develop an optimized outcome. It is an approach used to find the best possible behavior or path for challenges that involve sequential decision-making with multiple potential options at each step as commonly encountered in clinical situations such as the treatment of patients with sepsis [13]. A simplistic example is playing a game of chess or checkers. For each move there is a finite set of options that culminate in winning or losing the game. If the game was won, the program backtracks and assigns positive values to each move. If the outcome was a loss, negative values are placed on each move. Given thousands of games, dependent upon the board configuration, an optimal path at each move is developed.

"Unsupervised (learning) is most commonly used to refer to techniques that are just essentially doing clustering, there is no ground truth or gold standard with the data, but instead you're trying to organize it into meaningful groups of things and then upon inspection you might be able to infer something interesting about that ...."

Erik W. Brown

What is the evidence supporting the position that machine learning will play a disruptive role in clinical medicine? Reading the medical and lay literature it seems that new applications of AI and machine learning are trumpeted every week. Reports of the utility of machine learning algorithms to predict in-hospital mortality, length of stay and readmission [14]; classify distinct patterns of brain activity in individuals with psychiatric and cognitive disorders [15]; screen for asymptomatic left ventricular dysfunction using an office-based 12 lead electrocardiogram [16]; and diagnose genetic disorders [17] are but a few of the many applications being developed with direct relevance to the practice of clinical medicine. Are we reaching a technologic "tipping point" that will see machine learning applications permeate many aspects of clinical medicine or fade from relevance as did the much hyped IBM Watson for Oncology program? Already algorithms developed through machine learning have been shown to be equivalent or superior to existing approaches, including the performance of skilled physicians, in an array of clinically relevant situations. Let us look at some examples.

One application is detection of referable diabetic retinopathy—defined as moderate or worse diabetic retinopathy, diabetic macular edema, or both—in a population of patients with diabetes [18]. Using a data set of more than 100,000 retinal images from patients with diabetes, an algorithm was developed with 98% sensitivity and 94% specificity when set for a high sensitivity screening function and 90% sensitivity and 98% specificity when set for a high specificity diagnostic function. In a similar study using a smaller number of images to train the program, a sensitivity of 97% and specificity of 87% was reported [19]. Given the prevalence of diabetes and the morbidity associated with delayed diagnosis of referable diabetic retinopathy, the potential to automate, enhance and improve screening efficiency is promising. Encouragingly, prospective studies in "real world clinics" validated the technology and a commercial version of an imaging device combined with an algorithm made by IDx [20] has received FDA approval. This was the first such approval for an autonomous diabetic retinopathy screening system, meaning it was independent of a clinician.

Logically, as depicted in Fig. 8.1, the potential of machine learning to interpret retinal images extends to other ophthalmologic diagnoses including macular degeneration, glaucoma, and macular edema. While applying

FIG. 8.1 Algorithmic interpretation of retinal images can diagnose disease and predict risk for future events.

machine learning to retinal images to diagnose ophthalmologic diseases is essentially an extension of current practice, interestingly algorithms based on retinal images have also been developed to predict age, systolic blood pressure, gender, smoking status, and the 5 year risk for suffering a major adverse cardiovascular event (myocardial infarction or stroke) [21]. A promising area of study is the use of retinal images to provide a "window on the brain" thereby allowing the early diagnosis of neurodegenerative diseases [22].

Interpreting retinal images is an example where the number and geographic distribution of ophthalmologists with the requisite expertise is dwarfed by the need. A similar situation exists in dermatology. When applied to dermatologic photographs, training a deep neural network on almost 130,000 images of skin lesions with over 2000 different diagnoses, accuracy was demonstrated to be comparable to dermatologists' diagnosis. When compared with over 20 board-certified dermatologists, restricting the algorithm to differentiate between keratinocyte carcinomas, benign seborrheic keratosis, and malignant melanomas and benign nevi, the algorithm outperformed the average dermatologist [23]. The performance of such programs in "real world" clinics with a wide variety of individuals with different skin colors remains to be determined. With that caveat, the potential of mobile devices such as smart phones to greatly expand access to diagnostic care for dermatologic conditions is intriguing.

But the applications go beyond clinical images and traditional medical data. Hereditary angioedema is an autosomal dominant disease with a prevalence of 1 in 10,000 to 50,000 in the United States. Due to its rarity, it is often misdiagnosed for extended periods of time with attendant morbidity for the patient. Using a de-identified insurance claims database of over 170 million unique individuals treated during the time period from 2006 to 2014, patients with a diagnosis of hereditary angioedema were identified based on diagnostic coding and prescription of drugs specific for this condition. Machine learning was then employed to develop a claims database model of patients with hereditary angioedema and applied to the remaining patients. Over 5500 potentially undiagnosed individuals were identified. In an 11 month database update, nearly 900 of these potentially undiagnosed patients had additional health insurance claims, 14 of which now had codes for hereditary angioedema [24]. While this work is preliminary, it is intriguing to consider that databases such as those used for insurance claims might help identify individuals who potentially have rare diseases and alert their clinicians to consider the possibility.

"With the increasing accessibility of these algorithms or tools, comes the increased danger of unrecognized bugs. I've had students walk into my office and say, 'Look! We solved the problem, we have an accuracy of 99%,' and without fail we always find a bug. There are a lot of subtleties and issues that one can run into when setting up these problems and it's very easy to fool yourself into thinking that you have better results than reality. One has to be careful in setting up the experiment and the problem and really think about what it is that one is modeling.

For example, in terms of predicting who's at risk of infection and specifically infection with C. diff (Clostridioides difficile), we're actually predicting who's most likely to test positive for C. diff during the current hospital stay. Patients will only test positive if given the test, and will only be given the test if showing symptoms. Such subtle biases can present themselves in many different ways, so it's important to understand what it is you are actually learning. The model will only learn what you tell it to."

Jenna Wiens

We could review additional examples but I trust the reader will acknowledge that AI and machine learning will only grow in utilization. The question is the pace of introduction and scope of applications. Importantly, these are not simply technologic problems to be resolved but also involve regulatory and oversight policy decisions, portend economic disruption for clinical systems and practitioners, and have major sociologic implications.

"I think the hard part is understanding what the tool puts out, interpreting whatever it says. So it's not so much the model choice for your everyday physician, but an understanding of the strengths and weaknesses of the output of a model and when can I trust the model and under what circumstances is this model likely to have errors?"

H.V. Jagadish

While many of these considerations are beyond the control of individual physicians or clinical care delivery systems, there are important questions to consider prior to the adoption of machine learning generated algorithms in clinical care. Let us consider some examples that highlight potential pitfalls that may impact the clinical practice of the future.

Some developers are relatively agnostic to data quality and hold that, when it comes to machine learning, more data is better. They are confident that the programs will accurately identify "signal from noise." Traditional sources such as electronic health records, clinical trials, claims data, imaging data, and laboratory data, and nontraditional data such as that generated from geospatial and social media sources are all viewed as potentially contributing to clinically relevant algorithms. However, I would contend that the utility of an algorithm is predicated on the data set that is utilized to train the model and the quality of the learning database directly affects the quality of the resulting algorithm. The familiar adage of "garbage in, garbage out" is as applicable for algorithms derived from these mathematically complex programs as for simple traditional biostatistical analyses. The importance of the database is intuitively obvious. However determining its integrity is not always easy. Let us examine some examples of issues that might adversely affect algorithms generated through machine learning.

Example: Systematic error

A very simple example is developing a tool based on a database that includes clinical information only on men that is subsequently applied to men and women. Most sources of bias are not so apparent. Since machine learning algorithms are trained on existing databases, the bias that is inherent in any database requiring decision-making by humans may be magnified by the resulting algorithm.

As discussed by Mullainathan and Obermeyer [25], who sought to predict stroke risk during the week following an emergency room visit through an analysis of data from approximately 180,000 patients, systematic bias may lead to unanticipated findings. Potential predictors used to train the algorithm included demographic data and prior diagnoses present in the electronic health record over the year before the emergency department visit. Not surprisingly, they found that prior stroke had the strongest predictive value for short-term stroke risk. But this was followed by prior accidental injury, abnormal breast finding, cardiovascular disease history, colon cancer screening, and acute sinusitis. Clearly, with the exception of prior stroke and cardiovascular disease history, the other "predictors" are puzzling.

Understanding how the clinical database was constructed helps explain these non-intuitive findings.

To be included in the study information needed to exist in the patient's record. Thus, the following steps had to occur:

1. The patient must recognize the seriousness of their symptoms and seek care. If care is not sought obviously they cannot contribute to a clinical database.

2. The clinical personnel interacting with the patient must elicit the constellation of symptoms that raises the diagnostic possibility of stroke. This information, including risk factors and pertinent positive and negative history, must be entered into the database. If the possibility of stroke is not considered, the patient's information will not be entered in this database. Additionally, a wide array of factors and subjective decisions will determine what information is entered by the clinical personnel. Generally only data deemed pertinent to the diagnosis being entertained will be entered.

3. Appropriate tests to confirm or disconfirm the diagnostic hypothesis must be ordered. Here too the clinician's assumptions will determine what is required. If imaging studies are obtained, the radiologist's interpretation is key.

To have a diagnosis of stroke, the defined output, the patient needed to have sought medical care, be recognized as having stroke-like symptoms, and have appropriate tests to confirm the diagnosis and then have all the data accurately recorded in the medical record. While seemingly straightforward, there are many subjective decisions on the part of patients and clinicians that are inherent in existing clinical databases. In turn, this can lead to systemic bias and mismeasurement resulting in potential errors produced by machine learning tools.

Indeed, several of the "predictors" identified in this study are better thought of as predicting utilization of medical services. As such, these variables would emerge as predictors of any medical condition being studied. In this instance, the output is a conflation of stroke and utilization predictors and illustrates how the bias inherent in the development of the database is magnified by the algorithm, resulting in spurious predictors. While obvious in this situation, this is not always the case and misleading claims of new "risk factors" or new "biomarkers" may be the result.

A different sort of bias is introduced by "aged" training databases on which models are developed, validated and then applied not taking into consideration that the practice of medicine continues to evolve. The effort required to ensure that databases reflect the current clinical practices is considerable but necessary.

In response to a question regarding how to account for changing clinical practices: "The way in which a method is validated has important implications when interpreting model performance. Models can be validated on external data sets or on a randomly selected subset of the data that is held aside during training. However, this doesn't necessarily mimic how the model will be applied in practice and can lead to an overestimation of model performance. You can obtain good results, but that performance on the held portion of the data may not actually reflect how the model will perform in practice going forward, because,

exactly what you mentioned, things change. To address this, split the data temporally, instead of randomly. We train on data from say the first 5 years and then
test on the most recent year to see how robust the model is to changes over time."

Jenna Wiens

Even more problematic are the trade-offs between accuracy of the
model generated from the training database, its generalizability to new
data sets, and its intelligibility [12]. A model that accurately represents the
learning database is often highly complex. Ideally, such accuracy would
then be replicated when applied to new data sets. But this is rarely the
case. A balance between the complexity of the model and its usefulness for
new data is a critical developer-dependent decision which ultimately may
or may not be the correct one.

From a clinical and scientific perspective, highly accurate machine
learning-generated algorithms, using methods such as deep learning neural networks, generated from large data sets containing thousands of patients and attributes or features, are frequently unintelligible because the
underlying construct of the model is unknown. They may contain hundreds of layers leading to superb receiver operating curve (ROC) characteristics but no opportunity to understand the importance of a given
variable or feature on the part of the clinicians or scientists. Hence, the
acknowledgment that it is a "black box." It is worthwhile to remember the
adage—correlation is not causation—remains true for even the most sophisticated machine learning-generated algorithms. The tension between
accuracy and intelligibility is real and likely will remain so.

However, the potential for harm of black box models in high-stakes
decision-making increasingly is being recognized and the "necessity" of a
trade-off between accuracy and interpretability questioned. Indeed, for applications that have structured data with meaningful inputs often there is little or
no significant difference in performance between complex approaches such
as deep neural networks and simpler approaches such as logistic regression.
Consequently, it is argued that rather than trying to explain the findings from a
complex black box approach, the starting point should be creating models that
are interpretable. It is proposed that "no black box should be deployed when
there exists an interpretable model with the same level of performance" [26].

"(in regards to machine learning) That's my big worry, and it comes about
from my background as a physicist, there has to be causal relationship somewhere and many of these processes are so complex that you can't identify the
causal relationship, so it is garbage in, garbage out. I mean how can you have
any confidence in what it's telling you because there's not the fundamental
information in there on how these things relate to one another."

James Duderstadt

Example: Different programs, different results

Sometimes the explanation for unanticipated results is less obvious. One such study looked at predictors for cardiovascular events during a 10-year follow-up period. It was based on routine clinical data as of January 1, 2005 from over 378,000 patients in the United Kingdom who were then followed. At baseline, only individuals between the ages of 30 and 84 with no prior history of cardiovascular disease, inherited lipid disorder, or prescribed lipid-lowering drug, were enrolled. Four different machine learning algorithms (random forest, logistic regression, gradient boosting machines, and neural networks) were compared to the American College of Cardiology (ACC) algorithm to predict the first cardiovascular event over the ensuing 10 year follow-up period during which almost 25,000 cardiovascular events occurred.

To train the machine learning algorithms 75% of the study population was randomly assigned to a training cohort and the remaining 25% into a validation cohort. While the neural network algorithm performed best, each of the machine learning algorithms outperformed the ACC guideline in predicting which patients would have one of the 7400 first cardiovascular events that occurred in the validation cohort. That said, the sensitivity ranged from 62.7% for the ACC guideline to 67.5% for the neural network. The positive predictive value was 17.1% for the ACC guideline and 18.4% for the neural network. While an improvement, the machine learning algorithms outperformed expert opinion only modestly.

Interestingly, each of the four machine learning algorithms identified different sets of top 10 risk factors. All identified age, gender, ethnicity, smoking and social economic status (as measured by the Townsend Deprivation Index) among the top 10 risk factors for an initial cardiovascular event. Notably, a diagnosis of diabetes was not among the top 10 risk factors identified by any of the machine learning algorithms as measured by coefficient effect size. Among the unanticipated risk factors identified by one or more of the programs were atrial fibrillation, rheumatoid arthritis, chronic kidney disease, severe mental illness, and prescription of oral corticosteroids [27].

This study points out that machine learning algorithms are superior to expert judgment as represented by the ACC guideline and may identify previously unrecognized risk factors. However, whether the programs identified truly unique physiologically important "biomarkers" or risk factors or were the equivalent of "incidentalomas" found with new imaging technologies [28], remains to be determined.

In response to question as to which machine learning technique to employ: "At the end of the day, it is an engineering decision and one that should be informed by domain knowledge, but often one will try multiple different things.

For example, in predicting C. diff (Clostridioides difficile infection) one of the first things you think about is 'Well what are the types of data that we're dealing with?' Are they time varying, do they change over time? Are the data just a snapshot of the patient where the prediction task is - I just want to make a single prediction per patient per day? Or are the data natural language or free text, am I going to be processing physicians' notes? Or are the data images, do I have imaging data from the patients that I'd like to process?

First, understand what the data types are that you're going to be working with because for each of these different domains, or data types, different state-of-the-art techniques differ.

In addition to the data type, you need to think about the type of relationships among the input and the output. For example if the input is age, and you're trying to predict risk of infection, is it that in general the older you are, the more at risk you are of infection? Or is the relationship non-linear? Are both the very young and the very old at risk of infection? These relationships motivate the choice of pursuing a linear versus a non-linear model.

You can think about those things in advance and we do, but then very often we try both. For example, we may try random forest, a non-linear model, and then try something like a logistic regression, a linear model, and compare their performance. This can shed light on whether or not the underlying relationship between the input and the output is linear or non-linear."

Jenna Wiens

Example: Database corruption

Another consideration is the potential corruption of learning databases. As already noted, in detecting diabetic retinopathy, neural networks can be trained to outperform retinal specialists. However, when a high performance convolutional neural network trained on almost 500,000 diabetic retinopathy images was subsequently run with 50 additional diabetic retinopathy images subject to slight pixel modifications imperceptible to the expert human eye, the image-based algorithm tool's accuracy plummeted from 98% to 54%. Notably, a second hybrid lesion-based algorithm using multiple convolutional neural networks was minimally degraded [29]. This study raises the question as to the effect of randomly perturbed pixels in real-world patient images on algorithm development and reinforces the importance of judging different machine-based learning algorithms on their output. It also raises the possibility that the use of two or more different algorithms will minimize errors.

Beyond inadvertent pixel alterations, "adversarial attacks," where hackers subtly alter images, objects or sounds, leading to degradation of the machine learning algorithm are a valid concern. Cybersecurity has not been a major worry in the artificial intelligence machine learning community. A recent study of a subtly altered 3D printed turtle, an altered

3D printed baseball and a stop sign with stickers on it were identified by a neural network-generated algorithm as a rifle, an espresso and a speed limit sign respectively. All three objects were readily identified by people, with only the stickers on the stop sign being perceptible [30]. Not only is the need for robust cybersecurity highlighted by this study, so too is the need for research into the consequences of pixel alteration that might be imperceptible to the human eye, on the performance of sophisticated machine learning algorithms.

Example: Lack of transparency

Private companies are developing and marketing many machine learning algorithms. All too frequently, the learning and testing databases and the methodology employed are not disclosed for these proprietary products. The lack of transparency makes assessment of the resulting algorithm difficult.

One example, while based on simpler rules-based programming technology, is the widespread utilization of electrocardiogram (EKG) machines that provide interpretations concurrently with the printing of the EKG. While efforts to automate EKG analysis date to the 1950s, it did not become widespread until the 1970s, with the introduction of commercial machines. Now it is estimated that millions of EKGs are automatically interpreted annually, with the caveat to have them over-read by an expert. Likely, this advice is not routinely followed in many clinical settings around the globe and the automated interpretation is trusted to make clinical diagnoses and therapeutic decisions.

While the accuracy and reliability among different models from different manufacturers is high for normal tracings, incorrect readings are oftentimes generated for arrhythmias such as atrial fibrillation, conduction disorders including long QT syndrome, various pacemaker rhythms, and acute ST segment elevation myocardial infarction. Indeed, based on different proprietary interpretive programs, the false-positives range from 0% to 42% and false-negatives from 22% to 42% as compared to a panel of expert cardiologists when the machines are challenged to diagnose acute ST segment elevated myocardial infarction [31]. This is even after decades of refinements and enhancements of the programs upon which the tracing interpretations are based. The importance of sophisticated human oversight and judgment regarding machine-generated interpretations and the resulting diagnostic and therapeutic recommendations cannot be over emphasized.

This is illustrative of the complexities encountered in something as straightforward as interpreting an EKG. It is expected that with the advances in machine learning, these problems will be addressed if private companies, professional societies, healthcare systems and/or governmental agencies are willing to make the investments necessary to compile a well-curated tracing database large enough for learning as well as for validation testing. While the promise of machine learning is great for all digitized data, the experience with long-standing and sophisticated EKG technology is important to remember.

"There's a whole other side to this which is the regulatory aspect. It comes back to the explainability of the output of the system. If the underlying machine learning algorithm is a complete black box and is completely opaque and all you have is these different factors going in and then a recommendation coming out, how do you know whether or not you can rely on that if there's no explanation or link back to the evidence that was used?

So there are two parts to that; one is there is ongoing work on techniques to try and make that linkage and give you an explanation or at a minimum some sort of visualization, but the second, I think, is these machine learning algorithms need to be viewed very much like a drug that a life scientist or a pharmaceutical company is trying to evaluate."

Erik W. Brown

Unfortunately, with few exceptions, there are no open access databases of images, tracings, and similar digitally-generated data that contain the requisite hundreds of thousands to millions of high-quality inputs and well-documented outputs to allow training and testing and, very importantly, head-to-head comparisons to ascertain the most accurate and generalizable machine learning algorithms. The reality is that many of the marketed algorithms and those still under development do not share their strengths and weaknesses. While it is likely that each of these tools will be accompanied by the caveat that they require professional over-read and are meant to merely facilitate the work of clinicians, human nature being what it is, they will increasingly be relied upon as "accurate" interpretations.

Standards for database integrity, completeness and curation are necessary.

Much of the excitement surrounding the promise of machine learning has been due to the widespread introduction of electronic health records (EHR) and the allure of massive repositories of medical data that can be used for the development of diagnostic, predictive, and prognostic algorithms. However, the current state of record fragmentation, dependence upon clinicians' judgments as to what information to enter, unknown quantities of erroneous information entered either by accident or because of inaccurate interpretations of patients' histories and physical examination findings, and inaccurate recall by patients and their families, as well as the limitations imposed on the range of data to enter based on EHR developer decisions presents real challenges for the development of sophisticated machine learning based tools. While the potential to significantly enhance diagnostic accuracy is an admirable, and important, goal, whether or not machine learning will evolve to accurately diagnose the estimated 30,000+ known diseases remains an open question. Powerful machine learning programs exist and will become ever more sophisticated, but the limiting factor will be whether the requisite sophisticated databases for learning and validation testing will be created [32].

As science progresses and our understanding of the pathophysiologic basis of disease becomes increasingly more sophisticated, frequent revisions to the inputs and outputs of existing databases and relearning of algorithms will continue to be necessary. For example, as "genomic discoveries" that have been considered causal for specific diseases, are now recognized as polymorphisms that are found in the non-diseased population (but their prevalence had not been previously recognized), reference databases may need to be amended. Similarly, as the specificity of cancer diagnosis is enhanced as differing pathophysiologic mechanisms are discovered, databases and machine learning based algorithms to predict the likelihood of a given diagnosis or its prognosis will similarly need to be continuously updated. Ensuring that databases and their attendant curation are up-to-date requires the development of oversight, accreditation, and regulatory mechanisms that currently do not exist.

While being mindful of the admonition that technologic advances are not always progress, the application of machine learning developed tools across a rapidly expanding array of tasks will only proliferate. The impact on clinical medicine can only begin to be imagined.

Robotics

Building on progress made in sensors, computational power, miniaturization and other technologic advances, robots are increasingly deployed across a multitude of industries including clinical medicine. The first commercial use of robots was the 1961 introduction of Unimate by General Motors to the automobile assembly line. Since then, manufacturing has been transformed by robotics for repetitive tasks as quality and efficiency is enhanced. In many manufacturing processes, the human element has been replaced by the robot and the "operator" is now a specially trained computer engineer rather than a line worker. Robots are now deployed across a broad spectrum of applications in industry, military, search and rescue missions, deep-sea and space exploration, and even package delivery.

In the clinical setting, macrorobots such as those used in robotic-assisted surgery, have garnered the most attention. In this application, the surgeon's or operator's inputs are modified by the robot, allowing for greater precision and consistency and leading to the potential for enhanced patient outcomes. Through hand and/or foot controlled manipulators the robot filters and translates the clinician's movement into movements of the robotic arms and tools. Depending upon the robotic application, supporting staff prepare trocars and install the instruments and tools. As with all procedures, the skill of the clinician accounts for a considerable portion of the variability observed.

While the most widely used applications have been for laparoscopic procedures, robotic assisted neurosurgical and orthopedic procedures have also been deployed [33]. Some neurosurgical devices incorporate real-time MRI imaging information, a melding of technologies that enhances operative precision. The addition of real-time in addition to preoperative imaging to guide and direct the procedure will improve not only the efficiency and efficacy of anatomically complex procedures but will increasingly be incorporated into routine robotically assisted operations.

One of the potentials of robotic assisted surgery is enabling access to sophisticated surgical care in remote areas. The feasibility of transatlantic robot assisted telesurgery was demonstrated in a proof of concept laparoscopic remote robotic cholecystectomy in pigs with the surgeons controlling the operative console in New York and the remote surgical site in Strasbourg, France. Subsequently, a 68-year-old woman had a successful cholecystectomy done remotely by surgeons in New York while she was in Strasbourg. In this demonstration, surgeons were present at both sites and coordinated coagulation by voice command [34]. One of the constraints is the latency period between the transmission of operator commands, robotic action and the subsequent feedback to the operator. Research is ongoing into the effects of latency on cognitive workload, the operator's physiological and emotional responses and task performance [35]. To realize the potential of robotic telesurgery, understanding and enhancing the human operator function is arguably as or more important than developing ever more sophisticated surgical robots.

A probable next step will be to meld real-time imaging with sophisticated drivers to guide robotic surgery without a human operator. Fully autonomous surgeries for routine procedures such as appendectomies, cholecystectomies, and prosthetic joint replacements, are likely. However, potentially greater benefit may be had for relatively uncommon surgeries such as some neurosurgical and ophthalmologic procedures that require a high degree of precision and have a relative scarcity of surgeons proficient in the operations.

To date, robotic assisted surgical systems have been heavy, large, and expensive. They are limited by their mechanical linkages and require an external incision for access. But that is rapidly changing. Continuum robots, microrobots, and nanorobots hold great potential not just to enhance and replicate human procedural skills, but to enormously expand options across diagnostic and therapeutic applications.

Continuum robots, inspired by tentacles and snakes, can traverse and manipulate objects in complex, curvilinear environments. As they can be miniaturized and have far greater flexibility as compared to currently deployed rigid linked surgical robots their potential ranges from neurosurgical vascular approaches for intracerebral drug delivery and hemorrhage evacuation, to cardiac, vascular and urologic surgery [36]. Perhaps the most developed is natural orifice transluminal endoscopic surgery to access the abdomen, retroperitoneum, or mediastinum through an oral, vaginal, or rectal approach. Initial versions of the controls and continuum robotic instrumentation required to move this from research to clinical application are in production [37].

Microrobots and nanorobots differ fundamentally from macroscale robots. With their small size, they dramatically extend the array of potential clinical applications including targeted pharmacologic, radioactive, or hyperthermia therapeutic agent delivery; microsurgery such as removing endovascular fatty deposits, biopsy or ultrasound ablation of tissue; delivery of living cells for regenerative purposes to specific locations; using their physical properties to act as a scaffold for cellular regeneration, stenting, or occlusion; and diagnostically such as signaling the presence of a protein of interest, oxygen concentration in a specific region, or localizing internal bleeding sites.

The vasculature down to the capillaries, the central nervous system, the eye, the inner ear, and areas that are difficult to access with current technology are theoretically readily accessible. Diagnostic, regenerative, and therapeutic, the possibilities micro and nanorobots present only continue to grow [38].

Unlike macroscale robots that consist of sensors, actuators and programmable control units, building machines at the micrometer size precludes integrating all such components. Thus micro and nanoscale robots functionally differ from what is traditionally considered a robot. Lacking propulsion and steering systems, microrobots have been developed that are propelled and controlled via an external rotating magnetic field, light or near-infrared laser. And using external chemical gradients and proteins as "sensors" is ongoing research [39]. The technical problems posed by microrobots are magnified when moving at least an order of magnitude smaller to nanorobots - this is a focus of ongoing research.

While the attention paid to surgical robots has been great, the impact of robots on the clinical practice of medicine is just beginning. Technical innovations such as continuum robots that allow movement through complex spaces and the breakthroughs in miniaturization as applied to macro scale and now micro and nanoscale robots will expand their applications and capabilities. The diagnostic and therapeutic frontiers for robotic applications are just beginning to be explored [40].

The boundaries for clinical robotic applications extend beyond diagnosis and therapy. Their role as an adjunct to rehabilitation services for patients with neuromuscular impairment due to trauma and disease has been demonstrated in walking [41] and in upper extremity motor function recovery [42]. Multiple avenues of research on robot design for rehabilitation applications, as prosthetic devices, and assistive functions are being pursued. To name but a few, brain controlled prosthetic arms with the ability to grasp and provide feedback—haptics, brain controlled exoskeletons to allow ambulation, modular robotics designed for single joints such as a knee or ankle and "soft robotics" for individuals with physical limitations that preclude use of current rigid robots. It is anticipated that robots will be designed to replace or assist a wide range of human functioning.

The applications of robots across the spectrum of clinical care are only beginning to be recognized. Just as robotics fundamentally changed manufacturing processes, the likelihood of major disruptions in multiple areas of clinical practice and support is great. It is easy to anticipate that anesthesiology, critical care medicine, pathology, physical medicine and rehabilitation, and the procedural and surgical disciplines will be directly affected, but the ubiquitous presence of robots will transform care broadly.

Beyond the traditional boundaries of the clinical system, robotics is being viewed as an answer to the desire for aging individuals to remain independent and live in their own homes, minimizing the need for human personal assistants. While this includes technologic breakthroughs such as autonomous vehicles, we focus here on the role of home companion technologies. Studies have shown that the greatest needs that must be addressed are mobility, self-care, and social isolation [43]. Prototypes range from robotic platforms in the context of a smart home, utilizing a multitude of sensor arrays, to robots using in-home motion sensors. All employ machine learning to "learn" the person's daily patterns [44]. The ideal is to introduce the robot to the individual when they have minimal cognitive impairment, allowing the person to develop familiarity with the robot and the robot to learn the person's patterns and functioning. As cognitive impairment and physical frailty progress, the service robots might assist in activities as mundane as reminding to take medicines, making nutritional recommendations and encouraging physical activity, to providing lifting assistance and mobility. As needed, the robots would communicate with the clinical system regarding significant changes in the individual's function and, when a predetermined threshold is reached, the need for human clinical intervention, thus avoiding unnecessary visits to the physician or to the emergency room while also providing reassurance to the individual.

In Japan, human support and therapeutic robots range from "humanoids" being used to provide social interaction with elderly individuals to robotic "animals" being used as companions. While initial reports are encouraging, as yet unanswered is the question of the utility of robots to address the major problem of social isolation—especially for the elderly.

The applications of robots are myriad. In clinical medicine, they will become ubiquitous tools with a wide array of diagnostic, therapeutic and restorative uses. Robots will replace humans by providing care and assistance in many clinical settings, the community and the home. I anticipate that the disruptive role of robots in manufacturing will be repeated in medicine.

In this chapter I have highlighted only a few of the areas where innovations in technology will greatly impact clinical medicine. The convergence of engineering and medicine to enhance our understanding and to influence how care is provided is a powerful force for disruptive change. Advances in technology will enable fundamental changes in the practice of clinical medicine. Changes with the potential to greatly enhance access, monitor the response to therapy, automate repetitive tasks, interpret digitized data, and facilitate diagnosis; the possibilities are endless.

Whether this potential is realized in the near or distant future depends upon the rapidity with which necessary regulatory and licensure, reimbursement, and sociologic impediments are addressed. It will be the human factor not the technology that is the rate limiting step. Just as it has for the last 200 years, technologic innovation will shape the practice of medicine. Going forward, this will occur at an ever accelerating pace and in ways we are only beginning to imagine.

References

[1] Price AL, Spencer CCA, Connelly P. Progress and promise in understanding the genetic basis of common diseases. Proc R Soc B 2015;282:1684–93.

[2] Wangensteen OH, Peter ED, Nicoloff DM, Walder AI, Sosin H, Bernstein EF. Achieving "physiological gastrectomy" by gastric freezing: preliminary report of an experimental and clinical study. JAMA 1962;180(6):439–44.

[3] Hampton T. Critics of fibroid removal procedure question risks it may pose for women with undetected uterine cancer. JAMA 2014;311(9):891–3.

[4] Steinhubl SR, Waalen J, Edwards AM, Ariniello LM, Meha RR, Ebner GS, et al. Effect of a home-based wearable continuous ECG monitoring patch on detection of undiagnosed atrial fibrillation. JAMA 2018;320(2):146–55.

[5] Turakhia M, Perez M, on behalf of the Apple Heart Study Investigators. Results of a large-scale, app-based study to identify atrial fibrillation using a smartwatch: the Apple Heart Study. In: Presented at the 68th American College of Cardiology Scientific Session, New Orleans, Louisiana; March 16–18; 2019. Abstract 19-LB-20253.

[6] https://tricorder.xprize.org/.

[7] Noah B, Keller MS, Mosadeghi S, Stein L, Johl S, Delshad S, et al. Impact of remote patient monitoring on clinical outcomes: an updated meta-analysis of randomized controlled trials. NPJ Digit Med 2018;1:20172. https://doi.org/10.1038/s41746-017-0002-4.

[8] Cinar A. Multivariable adaptive artificial pancreas system in type 1 diabetes. Curr Diab Rep 2017;17:88–99. https://doi-org.proxy.lib.umich.edu/10.1007/s11892-017-0920-1.

[9] Li X, Dun J, Salins D, Zhou G, Zhou W, Schussler-Fiorenza Rose S, et al. Digital health: tracking physiomes and activity using wearable biosensors reveals useful health-related information. PLoS Biol 2017;15(1):1–30. https://doi.org/10.1371/journal.pbio.2001402.

[10] Gigerenzer G, Garcia-Retamero R. Cassandra's regret: the psychology of not wanting to know. Psychol Rev 2017;124(2):179–96.

[11] Trippel T, Weisse O, Xu W, Honeyman P, Fu K. WALNUT: waging doubt on the integrity of MEMS accelerometers with acoustic injection attacks. In: IEEE European symposium on security and privacy (EuroS&P); 2017. p. 3–18. https://doi.org/10.1109/EuroSP.2017.42.

[12] Deo RC. Machine learning in medicine. Circulation 2015;132(20):1920–30. https://doi.org/10.1161/CIRCULATIONAHA.115.001593.

[13] Komorowski M, Celi LA, Badawi O, Gordon AC, Faisal AA. The artificial intelligence clinician learns optimal treatment strategies for sepsis in intensive care. Nat Med 2018;24:1716–20.

[14] Topol E. High-performance medicine: the convergence of human and artificial intelligence. Nat Med 2019;25:44–56.

[15] Abbott B. Deeper learning. Machine learning makes new sense of psychiatric symptoms. Nat Med 2019;25:9–11.

[16] Attia Z, Kappa S, Lopez-Jimenez F, McKie PM, Ladewig DJ, Satam G, et al. Screening for cardiac contractile dysfunction using artificial intelligence-enabled electrocardiogram. Nat Med 2019;25:70–4.

[17] Gurovich Y, Hanani Y, Bar O, Nadav G, Fleischer N, Gelbman D, et al. Identifying facial phenotypes of genetic disorders using deep learning. Nat Med 2019;25:60–4.

[18] Gulshan V, Peng L, Coram M, Stumpe MC, Wu D, Narayanaswamy A, et al. Development and validation of a deep learning algorithm for detection of diabetic retinopathy in retinal fundus photographs. JAMA 2016;316(22):2402–10. https://doi.org/10.1001/jama.2016.17216.

[19] Abramoff MD, Lou Y, Erginay A, Clarida W, Amelon R, Folk JC, et al. Improved automated detection of diabetic retinopathy in a publicly available data set through integration of deep learning. Invest Ophthalmol Vis Sci 2016;57:5200–6. https://doi.org/10.1167/iovs.16-19964.

[20] https://www.eyediagnosis.net/ (Accessed March 20, 2019).

[21] Poplin R, Varadarajan AV, Blumer K, Liu Y, McConnell MV, Corrado GS, et al. Prediction of cardiovascular risk factors from retinal fundus photographs via deep learning. Nat Biomed Eng 2018;2:158–64.

[22] DeBue DC, Somfai GM, Koller A. Retinal microvascular network alterations: potential biomarkers of cerebrovascular and neural diseases. Am J Physiol Heart Circ Physiol 2017;312:H201–12.

[23] Esteva A, Kuprel B, Novoa RA, Ko J, Swetter SM, Blau HM, et al. Dermatologist-level classification of skin cancer with deep neural networks. Nature 2017;542:115–8.

[24] Kvancz DA, Sredzinski MN, Tadlock CG. Predictive analytics: a case study in machine-learning and claims databases. Am J Pharm Benefits 2016;8(6):214–9.

[25] Mullainathan S, Obermeyer Z. Does machine learning automate moral hazard and error? Am Econ Rev 2017;107(5):476–80. https://doi.org/10.1257/aer.p20171084.

[26] Rudin C. Stop explaining black box machine learning models for high-stakes decisions and use interpretable models instead. Nat Mach Intell 2019;1:206–15.

[27] Weng SF, Reps J, Kai J, Garibaldi JM, Qureshi N. Can machine learning improve cardiovascular risk prediction using routine clinical data? PLoS One 2017;12(4):e0174944. https://doi.org/10.1371/journal.pone.0174944.

[28] Berland LL, Silverman SG, Gore RM, Mayo-Smith WW, Megibow AJ, Yee J, et al. Managing incidental findings on abdominal CT: white paper of the ACR incidental findings committee. J Am Coll Radiol 2010;7:754–73.

[29] Shah A, Lynch SK, Niemeijer M, Amelon R, Clarida W, Folk JC, et al. Susceptibility to misdiagnosis of adversarial images by deep learning based retinal image analysis algorithms. In: IEEE 15th international symposium on biomedical imaging (ISBI 2018); 2018. p. 1454–7. https://doi.org/10.1109/ISBI.2018.8363846.

[30] Hutson M. Hackers easily fool artificial intelligences. Science 2018;361(6399):215. https://doi.org/10.1126/science.361.6399.215.

[31] Schlapfer J, Wellens HJ. Computer interpreted electrocardiograms benefits and limitations. J Am Coll Cardiol 2017;70:1183–92.

[32] Krumholz HM, Bourne PE, Kuntz RE, Paz HL, Terry SF, Waldstreicher J. Data acquisition, curation, and use for a continuously learning health system. Vital directions for health and healthcare initiative. National Academy of Medicine; September 19, 2016.

[33] Beasley RA. Medical robots: current systems and research directions. J Robot 2012;2012. 1–14. Article ID 401613. https://doi.org/10.1155/2012/401613.

[34] Marescaux J, Leroy J, Gagner M, Rubino F, Mutter D, Vix M, et al. Transatlantic robot assisted tele-surgery. Nature 2001;413:379–80.

[35] Yang E, Dorneich MC. The emotional, cognitive, physiological, and performance affects of variable time delay in robotic tele-operation. Int J Soc Robot 2017;9:491–508.

[36] Burgner-Kahrs J, Rucker DC, Choset H. Continuum robots for medical applications: a survey. IEEE Trans Robot 2015;31(6):1261–80.

[37] Lomanto D, Wijerathne S, Ho LKH, Phee LSJ. Flexible endoscopic robot. Minim Invasive Ther Allied Technol 2015;24(1):37–44. https://doi.org/10.3109/13645706.2014.996163.

[38] Nelson BJ, Kaliakatsos JK, Abbott JJ. Microrobots for minimally invasive medicine. Annu Rev Biomed Eng 2010;12:55–85.

[39] Palagi S, Fisher P. Bioinspired microrobots. Nat Rev Mater 2018;3:113–24.

[40] Huda MN, Yu H, Cang S. Robots for minimally invasive diagnosis and intervention. Robot Comput-Integr Manuf 2016;41:127–44.

[41] Esquenazi A, Talaty M, Jayaraman A. Powered exoskeletons for walk-in assistants and persons with central nervous system injuries: a narrative review. PM R 2017;9:46–62. https://doi.org/10.1016/j.pmrj.2016.07.534.

[42] Kim GY, Lim SY, Kim HJ, Lee BJ, Seo SC, Cho KH, et al. Is robot-assisted therapy effective in upper extremity recovery in early stage stroke?—a systematic literature review. J Phys Ther Sci 2017;29(6):1108–12. https://doi.org/10.1589/jpts.29.1108.

[43] Bedaf S, Gelderblom GJ, de Witte L. Overview and categorization of robots supporting independent living of elderly people: what activities do they support and how far have they developed. Assist Technol 2015;27:88–100.

[44] Fiorini L, Cavallo F, Dario P, Eavis A, Caleb-Solley P. Unsupervised machine learning for developing personalized behavior models using activity data. Sensors 2017;17:1034–51.

Additional resources

- AI Index. The mission of AI Index is to: ground the conversation about AI in data. It tracks, collates, distills and visualizes data relating to artificial intelligence. https://aiindex.org/.
- Wachter RM. The digital doctor: hope, hype, and harm at the dawn of medicine's computer age. New York, NY: McGraw Hill; 2015.
- Topol E. The creative destruction of medicine. How the digital revolution will create better healthcare. New York, NY: Basic Books; 2012.
- Rajkomar A, Dean J, Kohane I. Machine learning in medicine. N Engl J Med 2019;380(14):1347–58.

128

8. Technology and computing

- Tcheng JE, Bakken S, Bates DW, Bonner H III, Gandhi TK, Josephs M, et al., editors. Optimizing strategies for clinical decision support: summary of a meeting series. Washington, DC: National Academy of Medicine; 2017.
- National Academies of Sciences, Engineering, and Medicine. Artificial intelligence applications for older adults and people with disabilities: balancing safety and autonomy: proceedings of a workshop in brief. Washington, DC: The National Academies Press; 2019. https://doi.org/10.17226/25427.
- National Academies of Sciences, Engineering, and Medicine. Harnessing mobile devices for nervous system disorders: proceedings of a workshop. Washington, DC: The National Academies Press; 2018. https://doi.org/10.17226/25274.
- National Research Council. Frontiers in massive data analysis. Washington, DC: The National Academies Press; 2013. https://doi.org/10.17226/18374.
- National Academies of Sciences, Engineering, and Medicine. The promise of assistive technology to enhance activity and work participation. Washington, DC: The National Academies Press; 2017. https://doi.org/10.17226/24740.

CHAPTER

9

Microbiota and the microbiome

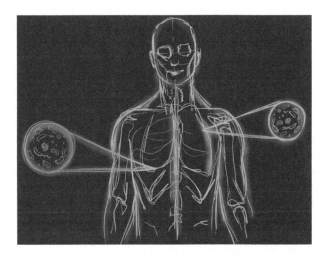

Our understanding of the microorganisms that live on or within our tissues and in our biofluids has exploded within the last 10 years. We now know that bacteria, archaea, fungi, protists, and viruses are ubiquitous and their cell and gene number exceeds that of the human host they populate. It is estimated that every person is in a generally symbiotic relationship with between 10 and 100 trillion microbial cells. This complex ecosystem is called the microbiota. Although estimates vary, it may be composed of a thousand or more different species, many of which remain to be identified. As technologic advances enabled rapid, affordable, and sophisticated sequencing of the microbiome, the genetic material of microbial ecosystems, there have been thousands of published articles showing correlations between alterations in the microbiomes of various organs and a multitude of different diseases. Generally the literature will reference the microbiome, as it is the genes that are being identified rather than the specific microbial species.

Implementing Biomedical Innovations into Health, Education, and Practice
https://doi.org/10.1016/B978-0-12-819620-5.00009-6

"We're just discovering what microbiome elements are associated with what. But I really believe it's going to be like the revolution that we saw with antibiotics. I'd predict 30 to 50 years from now, things that manipulate the microbiome are going to be as commonplace as our portfolio of antibiotics, if not more common."

Joseph Kolars

As illustrated by the graphic in Fig. 9.1, a wide variety of bacteria inhabit one organ, the skin, and the classes of bacteria vary by location. Similar complex ecosystems have been identified in all organs studied. Furthermore, studies have shown that individuals have unique microbiota although family members or others with whom they live in close proximity show similarities. It is a dynamic ecosystem. For example, dietary changes will rapidly alter the gut microbiome as will environmental factors such as geographic location and behaviors such as smoking.

While disease-causing pathogenic bacteria, viruses, and fungi have been identified and studied for decades, we now recognize that many organisms, initially thought to be commensal, can influence human health and disease. For example the intestinal microbiome, the most widely studied, plays important roles in the maturation and education of the immune system, regulates intestinal endocrine functions, affects brain function, influences bone density, and biosynthesizes vitamins, neurotransmitters and other compounds with yet to be determined functions, in addition to its role in energy production through food digestion [1].

Despite these advances, our understanding of the human microbiota is limited. As we have come to appreciate the sheer magnitude, diversity of species, and metabolic potential of the microbiota inhabiting each of us, the possibility that it plays a central role in our health and disease states is appealing. Whether specific species or clusters of species within the microbiota will prove to be the causative etiology for a wide array of diseases, be found to play a "supporting role" in the development and/or severity of disease states or will be found of importance in only a limited number of diseases remains to be determined.

There are studies linking alterations in the gut microbiome to risks for: atherosclerosis, obesity, type 1 and type 2 diabetes, hypertension, cancers including colon and pancreatic, a host of inflammatory and autoimmune diseases, neurologic diseases and psychiatric disorders. However, the vast majority of these have simply been correlations, without proven causal mechanisms or pathways. Furthermore, while we can envision that manipulation of the microbiota/microbiome will be of therapeutic benefit, the extent to which "microbiome therapy" will become a major part of clinicians' therapeutic options for treating disease or maintaining health remains conjecture at this point.

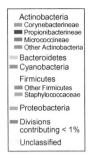

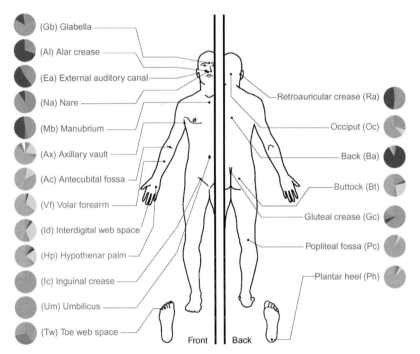

FIG. 9.1 Graphic depicting the human skin microbiota, with relative prevalence of various classes of bacteria. *Reproduced from Wikimedia commons. Credit: Darryl Leja, NHGRI—Public Domain,* https://commons.wikimedia.org/w/index.php?curid=29534265.

Recognizing the complexity of each of our microbiomes and how little we know about their deterministic or facilitative roles in health and disease, it is difficult to gauge the likelihood that knowledge of and manipulation of our microbiota will be a major disruptor in mid-century clinical practice. That said, let us look at some specific areas where tantalizing clues are emerging as to the role of the microbiota/microbiome in health and disease.

Neurologic and psychiatric disease

Gastrointestinal complaints such as constipation, diarrhea, and abdominal pain have long been recognized as a feature of many psychiatric and neurologic disorders. Research has now shown that the gut microbiome can influence brain function and behavior, and vice versa, it is a bidirectional effect. This observation has led to what is termed the "gut-brain axis." It is thought that vagus nerve stimulation; production of neuromodulatory metabolites including short-chain fatty acids, neurotransmitters such as serotonin, GABA and dopamine and hormones such as catecholamine; and immune mediated inflammatory pathways are involved. While the majority of the studies linking alterations of the intestinal microbiome with neurologic and psychiatric diseases have been conducted in animals, they provide intriguing findings that gut bacteria can influence anxiety, stress, depression, and social behavior; and that the development of and severity of animal models of multiple sclerosis, Parkinson's disease, ALS and Alzheimer's disease are influenced by the gut microbiome. Studies in patients with schizophrenia, bipolar disorder, autism spectrum disorder, relapsing remitting onset of multiple sclerosis and Parkinson's disease have all shown differences in their microbiome, termed dysbiosis [2]. The complex interactions mediated by the gut-brain axis are the focus of much research activity and are just beginning to be elucidated.

Beyond looking for differences in the microbiomes between patients and normal individuals, there have been small interventional studies. Healthy women with no gastrointestinal or psychiatric symptoms were randomly assigned to receive twice daily for 4 weeks a fermented milk product with probiotics (yogurt) or a non-fermented milk product, or no intervention. Functional magnetic resonance imaging (fMRI) scanning to measure resting and brain response to an emotional faces attention task was done before and at completion of the 4 week intervention. The women receiving the yogurt demonstrated reduced generalized brain response to negative images following intervention whereas there was an increased response in the non-fermented milk product group and no change in the group that did not receive intervention [3]. A similar study of oral probiotic administration over 30 days showed alleviation of anxiety and depression as measured by the Hospital Anxiety and Depression Scale and decreased urinary free cortisol [4]. Whether probiotics have an anxiolytic effect due to their effect on the gut microbiome remains to be resolved.

Atherosclerosis

Evidence is accumulating that links the gut microbiome to endothelial damage and plaque formation through trimethyl amine N-oxide (TMAO).

A volatile gas, trimethylamine (TMA), is a byproduct of the gut microbiota metabolizing choline and L-carnitine in food. As TMA passes through the portal circulation into the liver, it is transformed into TMAO by the hepatic enzyme, flavin monooxygenase 3. TMAO induces inflammation, increases platelet activity and thrombosis, and formation of foam cells. A review of prospective human studies showed that high levels of TMAO and its precursors significantly increased the relative risk for major adverse cardiovascular events and death, independent of traditional risk factors [5]. Beyond this direct effect it appears that plaque progression and rupture is accelerated due to low grade gut inflammation and passage of bacteria and bacterial products into the circulation secondary to increased gut permeability and disruption of tight junction proteins. And finally, the gut microbiome is linked to the development of risk factors including obesity, type 2 diabetes, cholesterol metabolism, and hypertension [6].

Obesity

Comparison of lean and obese mice shows differences in the relative abundance in the cecal microbiota of two bacterial divisions, Bacteroidetes and Firmicutes, and transplantation of feces from obese mice into germ-free normal mice induced significantly greater increases in body fat over 2 weeks as compared to fecal transplantation from lean mice. This has spurred tremendous interest in the gut microbiome and its role in obesity [7]. Subsequently, fecal transplantation from lean mice to obese mice induced weight loss without a change in caloric ingestion. Multiple theories as to the basis for this observation have been proposed. More efficient digestion and energy conversion of ingested food, effects of bacterial metabolites on brain satiety centers, and low-grade systemic inflammation leading to impaired insulin sensitivity are but some of the mechanisms that have been postulated.

Initial human studies showed a similar lower ratio in obese individuals of Bacteroidetes to Firmicutes, However, subsequent large human studies, including the Human Microbiome Project, have failed to replicate these findings [8]. Much more work needs to be done before we can speak with confidence about the findings and their implications.

When considering weight loss and the gut microbiota in obese individuals, there are several interesting observations.

- Dietary changes associated with weight reduction have been shown to alter the gut microbiome.
- When comparing obese individuals who successfully lost at least 5% of their body weight at 3 months to those who were unsuccessful, a gut microbiome with increased capability for carbohydrate metabolism was found in the unsuccessful individuals.

- A double-blind randomized placebo-controlled study of a lactobacillus probiotic intervention showed positive effects on weight reduction.
- Bariatric surgery rapidly results in significant long-lasting alterations in the composition of the gut microbiome. These changes are temporally related to the early weight loss that is observed, disproportionate to the decrease in caloric intake. Potentially, alterations in bile acid metabolism directly influence the gut microbiome changes.
- Studies of human fecal transplantation are largely in patients with persistent Clostridioides difficile infection with only anecdotal comments regarding the effect on weight.

Much of the evidence regarding the role of the gut microbiota in obesity stems from studies of animal models, primarily rodents. As such, it is critical to remember that rodents have a markedly different gut microbiome and immune system. Inferences derived from these animal studies may not translate directly to humans. While intriguing, the role of the gut microbiome in the development of obesity in humans and how manipulation of that ecosystem will affect obese individuals remains to be determined [9].

Type 1 and type 2 diabetes mellitus

Mechanisms that are considered to play causal roles in other chronic diseases are similarly thought to be at least contributing to diabetes. The association between obesity and type 2 diabetes is strong. Gut inflammation and microbiota metabolic products leading to insulin resistance and increased production of ghrelin have been linked to type 2 diabetes as well [10].

Few studies have looked at intentional manipulations of the gut microbiome and its effect in humans. A study of patients diagnosed with type 2 diabetes were randomized to receive either usual care or a high-fiber diet composed of whole grains, traditional Chinese medicinal foods, and prebiotics. The daily caloric and macronutrient intakes were the same in the two diets. The high-fiber diet increased the availability of non-digestible but fermentable carbohydrates leading to increased short-chain fatty acid production by the gut microbiome. Significant improvements in hemoglobin A1c were observed in the high-fiber diet cohort [11].

In a prospective cohort study of infants at high risk for the development of type 1 diabetes mellitus due to their HLA genotype, blood samples were obtained every 3 months through 4 years of age and every 6 months thereafter to determine whether persistent islet cell autoimmunity developed. Details of their feeding, including probiotic supplementation and formula use, were monitored from birth. The children with a DR3/4 genotype receiving probiotic supplementation from ages 0 to 27 days had a 60% decrease in islet autoimmunity when compared to later probiotic supplementation within the first year a life or no supplementation. It is postulated that probiotics received via fortified infant formula or dietary supplements may alter the development of their immune systems and the development of islet autoimmunity [12].

As with other chronic diseases, the implications, based on animal studies, that the gut microbiome plays a role in the development of type 2 diabetes and to a lesser extent type 1 diabetes mellitus is intriguing. Confirmation with human studies awaits, as does the potential for therapeutic interventions based upon manipulation of the gut microbiota.

Not unexpectedly, the microbiome has been postulated to play a role in a host of other diseases including connective tissue diseases, vasculitides, colorectal and pancreatic cancers, chronic liver disease, chronic obstructive pulmonary disease, asthma, eczema, to name but a few. A recurring theme that underlies many chronic diseases is evidence of low-grade systemic inflammation. In animal models, alterations in the gut microbiome have been shown to lead to such systemic inflammation. In humans, there is plausible evidence for such a relationship. Could it be that the multitude of epidemiologic studies linking dietary patterns to chronic disease are mediated through the gut microbiota?

While much of the focus has been on the gut, what is the effect of environment and behavior on the microbiota of other organs? It is intriguing to consider the possibility that the microbiome is contributing to or is responsible for a multitude of chronic diseases. We are just beginning to unravel the complex web of interactions mediated by what is sometimes characterized as the largest endocrine organ, our microbiome, and its effect on a multitude of organs and functions.

As with so many studies in the life sciences, the caution that correlation is not causation is especially true for studies of the microbiome. The fact that environmental changes, including, for the gut dietary changes, can rapidly alter the microbiome composition is well-established. Whether observed changes in the microbiome with different disease states are simply secondary to the disease or actually are playing a primary or secondary causal role must be established.

Microbiota and pharmaceutical interactions

As the microbiome increasingly is being identified as playing a role in disease causation, its importance in modifying therapeutic responses is just beginning to be appreciated. For example, the gut microbiome appears to influence the efficacy of cancer immunotherapy. Patients with altered gut microbiomes due to antibiotic ingestion had reduced progression-free and overall survival compared to patients who had not taken antibiotics within the 2 months before and 1 month after the first administration of anti-PD-1 immune checkpoint inhibitors [13, 14]. While it has been known for decades that antibiotics alter the gut microbiota, the observation that this in turn impacts the efficacy of chemotherapy raises the question as to how many other such interactions will be discovered. Moreover, recently it has been reported that about 25% of all marketed drugs inhibit the growth of at least one strain of gut bacteria in vitro [15]. The consequences of altered microbiota ecosystems on the efficacy of pharmacologic therapy and the effects of pharmacologic agents on the microbiota are just beginning to be explored. The implications of such interactions for clinical therapeutics remains to be determined.

While intriguing, the contributions of the microbiome to health and disease remain to be determined for the majority of conditions in which microbiome alterations have been implicated. It is likely that much will be elucidated in the next several decades and will have a profound effect on the practice of clinical medicine in mid-century. We must prepare our students accordingly.

References

[1] Lynch SV, Pederson O. The human intestinal microbiome in health and disease. N Engl J Med 2016;375:2369–79.
[2] Tremlett H, Bauer KC, Appel-Cresswell S, Finlay BB, Waubant E. Gut microbiome in human neurologic disease: a review. Ann Neurol 2017;81:369–82.
[3] Tillisch K, Labus J, Kilpatrick L, Jiang Z, Stains J, Ebrat B, et al. Consumption of fermented milk product with probiotic modulates brain activity. Gastroenterology 2013;144:1394–401.
[4] Messaoudi M, Violle N, Bisson JF, Desor D, Javelot H, Rougeot C. Beneficial psychological effects of a probiotic formulation (*Lactobacillus helveticus* R0052 and *Bifidobacterium longum* R0175) in healthy human volunteers. Gut Microbes 2011;2:256–61.
[5] Heianza Y, Wenjie M, Manson JE, Rexrode KM, Qi L. Gut microbiota metabolites and risk of major adverse cardiovascular disease events and death: a systematic review and meta-analysis of prospective studies. J Am Heart Assoc 2017;6(7):e004947. https://doi.org/10.1161/JAHA.116.004947.
[6] Komaroff AL. The microbiome and risk for atherosclerosis. JAMA 2018;319(23):2381–2.
[7] Turnbaugh PJ, Ley RE, Mahowald MA, Magrini V, Mardis ER, Gordon JI. An obesity associated gut microbiome with increased capacity for energy harvest. Nature 2006;444:1027–131.
[8] Finucane MM, Sharpton TJ, Laurent TJ, Poilard KS. A taxonomic signature of obesity in the microbiome? Getting to the guts of the matter. PLoS One 2014;9:e84689. https://doi.org/10.1371/journal.pone.0084689.
[9] Bouter KE, van Raalte DH, Groen AK, Nieudorp M. Role of the gut microbiome in the pathogenesis of obesity and obesity-related metabolic dysfunction. Gastroenterology 2017;152:1671–8.
[10] Komaroff AL. The microbiome and risk for obesity and diabetes. JAMA 2017;317(4):355–6.

[11] Zhao L, Zhang F, Ding X, Wu G, Lam YY, Wang X, et al. Gut bacteria selectively promoted by dietary fibers alleviate type 2 diabetes. Science 2018;359:1151–6.

[12] Uusitalo U, Liu X, Yang J, Aronsson CA, Hummel S, Butterworth M, et al. Association of early exposure probiotics and islet autoimmunity in the TEDDY study. JAMA Pediatr 2016;170(1):20–8.

[13] Routy B, Le Chatelier E, Derosa L, Duong CPM, Alou MT, Daillèr R, et al. Gut microbiome influences efficacy of PD-1-based immunotherapy against epithelial tumors. Science 2017;359(6371):91–7. https://doi.org/10.1126/science.aan3706.

[14] Gopalakrishnan V, Spencer CN, Nezi L, Reuben A, Andrews MC, Karpinets TV, et al. Gut microbiome modulates response to anti-PD-1 immunotherapy in melanoma patients. Science 2017;359(6371):97–103. https://doi.org/10.1126/science.aan4236.

[15] Maier L, Pruteanu M, Kuhn M, Zeller G, Telzerow A, Anderson EE, et al. Extensive impact of non-antibiotic drugs on human gut bacteria. Nature 2018;555:623–8. https://doi.org/10.1038/nature 25979.

Additional resource

- National Academies of Sciences, Engineering, and Medicine. Microbiomes of the built environment: a research agenda for indoor microbiology, human health, and buildings. Washington, DC: The National Academies Press; 2017. https://doi.org/10.17226/23647.

CHAPTER

10

Genetics and molecular biology

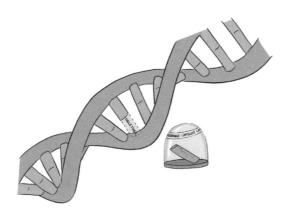

The announcement of the draft sequence of the human genome on June 26, 2000 and completion of the Human Genome Project in 2003 was heralded as a major milestone in our understanding of health and disease. The surrounding "euphoria" raised expectations that the genetic basis for many maladies would now be identified and amenable to therapy at their fundamental genetic levels. Indeed as is registered in *Online Mendelian Inheritance in Man* (OMIM®) which is a "continuously updated catalog of human genes and genetic disorders and traits, with particular focus on the molecular relationship between genetic variation and phenotypic expression" there are almost 25,000 entries of disease associated genetic variants and the number increases daily [1].

However, the expectation that disease risk would be readily linked to genomic variants, with a few notable exceptions, has proven elusive. Despite thousands of studies the literature is frequently conflicting as to the pathogenicity of a given variant. Claims of pathogenicity associated with a genetic variant are found to be overstated and subsequent large-scale exome studies have led to a downgrade reclassification of hundreds of initially claimed to be pathogenic genomic variants for a range of diseases [2].

The importance of sequencing the genomes from ethnically diverse populations is now recognized. In a study of risk stratification for hypertrophic cardiomyopathy, it was found that variants that had been considered pathogenic based on a white American control cohort are actually benign variants more commonly found in black Americans. This had led to the initial misclassification of some African-Americans as having pathogenic variants [3]. Genomic misclassifications and the resulting misdiagnosis due to biased samplings of a control cohort are increasingly recognized. This problem exists in multiple disease categories. Notably, as the numbers of individuals participating in genetic testing for hereditary cancer susceptibility has expanded beyond a predominantly Caucasian population it has also been found that reclassification rates are ancestry dependent [4]. It is suggested that misclassification of genomic variants as pathogenic may account, at least in part, for the "genotype-positive—phenotype-negative" phenomenon. Until massive reference control cohorts that appropriately include individuals from different ancestries are developed, ancestry matched controls are optimal. However this may not be feasible for some ancestral groups and often the ancestry data for the control cohorts may not be available or even known.

Nevertheless, progress continues. As databases have grown, it has become possible to reclassify variants that previously had unknown significance (VUS) and confirm earlier classifications. In a retrospective cohort study of 1.45 million individuals who were genetically tested for hereditary cancer at a single commercial laboratory from 2006 to 2018, 6.4% of unique variants were subsequently reclassified. Very importantly, reclassification to a different category was rare among variants classified initially as pathogenic or likely pathogenic (0.7%) or benign or likely benign (0.25%). Variants of uncertain significance were much more likely to be reclassified (7.7%) with 91.2% downgraded to benign or likely benign [5]. Our ability to classify genetic variants as to their pathogenicity will continue to be refined.

Assessing the risk of adults for disease is just one application of genetic screening. A recent report from the BabySeq Project highlights the potential importance of newborn genomic sequencing. One hundred twenty seven healthy newborns and thirty two ill newborns in a neonatal intensive care unit were enrolled. Pathogenic or likely pathogenic variants in dominantly, or bi-alles in recessively, inherited diseases, carrier status, and pharmacogenomic findings of relevance to the pediatric population were reported. 15/159 genomic sequences revealed unanticipated risks for childhood onset disease (10 of which were in healthy newborns) and 3.5% also had an actionable adult onset disease risk. 140/159 genomes revealed a carrier state for a recessive disorder and 8/159 had genomic findings that might influence future pharmaceutical prescribing decisions [6]. The prevalence of potentially important genomic findings in these newborns is notable. Even after further investigation, family history identified a diseased relative who might share the pathogenic variant in only three instances.

As exemplified in the BabySeq study it is argued that routine implementation of newborn genetic sequencing might identify risk for diseases from childhood through adult life, facilitate rapid diagnosis for sick newborns and children, inform future pharmaceutical choices and dosing (pharmacogenomics), and determine carrier status for recessive diseases that might impact reproductive decisions. Beyond the considerable operational challenges associated with implementation of massive genomic screening and subsequent data analysis, there are a host of issues to be resolved. With the exception of a relatively few rare diseases it is likely that a multitude of environmental and gene-gene interactions will play important roles in determining whether a given variant actually leads to disease. Parents will be seeking information as to the severity and timing of onset of disease associated with their newborn's findings. Recognizing that our understanding is not sufficient to answer many of the questions that parents would pose, how do clinicians appropriately communicate to parents the risks that genomic sequencing reveals in their newborns? As medical educators how do we teach our students and residents to have these conversations?

Despite the variable evidence linking genomic variants and disease, the NIH Genetic Testing Registry as of March 25, 2019 contained information from 519 laboratories on 59,095 tests for 11,494 conditions involving 18,595 genes [7]. To date, the norm has been to report genomic variants as pathogenic, likely pathogenic, variants of unknown significance, likely benign or benign. But arguably just as important is the reporting of protective variants. Understanding this gene-gene interaction and its effect on disease development will be necessary to refine risk stratification for a given individual [8].

One example is risk for coronary artery disease (CAD). *PCSK9* loss-of-function variants are associated with a decreased risk for CAD and very low levels of low-density lipoprotein cholesterol. Results from the MyCode Community Health Initiative at Geisinger Health System of 50,726 individuals who underwent exome sequencing for familial hypercholesterolemia based on pathogenic variants in three genes, *LDLR*, *APOB*, and *PCSK9* were recently reported. Two hundred twenty nine individuals were identified as heterozygous carriers of pathogenic variants for familial hypercholesterolemia. Overall these individuals had significantly increased odds of general and premature CAD. However, not everyone had clinical evidence of CAD or even elevated LDL-C. Potentially, this is at least partially explained by a protective *PCSK9* variant [9].

We are only beginning to understand the complexities involved in the phenotypic development of disease in individuals with identified genetic risks. Despite the increased knowledge of genetic alterations in disease development, many questions remain, among them:

- What pathway(s) does a given gene regulate and/or participate in?
- What turns on a pathway that results in a malignancy in adults but is required for normal development?
- What are the implications for pathway interactions of various genetic variants?
- What are the environmental and behavioral factors that lead to epigenetic changes and gene regulation?

But the real goal is to be able to intervene to prevent the onset of disease. If this becomes a reality for even a subset of diseases it has far-reaching implications for the role of clinicians in medical practice and how we prepare future physicians for such a reality.

Gene therapy

The discovery that some rare, often fatal at a young age or severely disabling, inheritable diseases were due to a single gene mutation led to the concept of being able to repair or replace the defective gene, gene therapy, almost half a century ago. This idea that some inherited diseases could be cured or at least the disease course permanently altered by introducing a functional gene has intrigued clinicians and the public, and spurred basic science and translational research ever since.

Early in the development of gene therapy, viral vectors were recognized as a means to introduce genes—gene adding—to target cells. While there were promising advances, there were also unanticipated toxicities including development of malignancies, host immunologic inflammatory responses and patient death.

As has often been the experience with revolutionary clinical therapies, initial optimism was tempered by the reality of clinical trials but the setbacks pointed to areas where further research was needed to understand the basic mechanisms. That led to improved vectors, increased knowledge of host immunologic responses, further work to minimize the potential for proto-oncogene activation, and the development of alternative delivery mechanisms that have continued to move the field forward [10]. This is an important lesson for us to remember. The translation of insights gained through basic research into clinical application is an unpredictable process that is often more complicated than initially envisioned.

A parallel line of development is based on ex vivo genetic engineering. Using engineered hematopoietic stem cells that have been induced to produce missing enzymes and proteins, trials treating hemoglobinopathies, metabolic disorders and lysosomal storage diseases have been promising.

Even more interestingly, this approach has the potential to revolutionize the treatment of thalassemia and sickle cell disease, inherited disorders that affect millions of people worldwide.

While initially envisioned for monogenetic inherited diseases, gene therapy is increasingly focused on therapeutic approaches for cancers and complex multigenic diseases. A potentially major advance has been cancer immunotherapy based on chimeric antigen receptor (CAR) engineered T cells that has resulted in new FDA approved approaches to the treatment of leukemia and lymphoma and has stimulated ongoing work focused on other malignancies. However, serious systemic toxicities including cytokine release syndrome and neurotoxicities have been reported, highlighting the need for deliberate clinical studies and additional research to understand the pathophysiologic mechanisms underlying the untoward side effects [11].

As defined by the federal oversight agency, the US Food and Drug Administration (FDA) Center for Biologics Evaluation and Research (CBER), "Human gene therapy seeks to modify or manipulate the expression of a gene or to alter the biological properties of living cells for therapeutic use." Gene therapy has now matured to the point that there are over 3800 gene therapy trials, 2900 of which are registered as interventional, on ClinicalTrials.gov. When looking specifically at gene editing, there are 26 such trials registered [12]. To date, clinical trials have shown benefits in a wide array of human conditions. FDA approval has been given for an adeno-associated virus mediated gene therapy treatment for inherited retinal dystrophy and European Medicines Agency (EMA) approval for a lipoprotein lipase deficiency therapy. It is anticipated that regulatory approval will be given for an ever increasing number of gene therapy based therapeutic interventions.

Gene editing

The ability to target gene editing in situ has been a dramatic recent advance. Perhaps the best known gene editing technology is CRISPR— clustered regularly interspaced short palindromic repeats. CRISPR and similar gene editing technologies have provided biomedical scientists a much easier, more rapid and less expensive means to edit genes and allowed for greater understanding of how the genome operates. The potential to precisely correct human diseases, ranging from single to multigenic pathologies, has drawn significant attention. Among the numerous potentials is treatment of muscular dystrophy and solid tumors, editing of genetic variants that carry a risk for disease prior to disease onset, and germline editing to eliminate disease potential in future generations [13]. However a multitude of safety and feasibility issues such as "off target" mutations [14], immunogenicity, and target tissue delivery remain to be resolved. Perhaps the greatest hurdle is not biologic in nature but rather involves the social and ethical issues associated with germline editing.

The 2015 Chinese research publication using CRISPR-Cas9 to modify "nonviable" pre-implementation human embryos resulted in discussions, conferences and a publication from the US National Academies of Sciences, Engineering and Medicine establishing principles of governance and oversight for human genome editing [15]. However, the announcement in November 2018 that a human embryo had successfully been edited using CRISPR/Cas9 and that gene edited embryos resulted in the birth of twins resulted in shockwaves through the research and academic communities [16]. Despite widespread statements of concern that such clinical applications are premature, it is apparent that germline gene editing is feasible and that our scientific and technologic capabilities have outdistanced our ethical deliberations. As Director General of the World Health Organization (WHO) Tedros Adhanom Ghebreyesus stated in a March 19, 2019 press release announcing the creation of a global registry to monitor gene editing research in humans "gene editing holds incredible promise for health, but it also poses some risks, both ethically and medically." The WHO advisory panel did not call for a moratorium on human germ line editing. In contrast, 18 prominent scientists and bioethicists from seven countries called for a "fixed period during which no clinical uses of germline editing whatsoever are allowed" [17]. There are parallels to the development of recombinant DNA and the process of self-governance that the scientific community exerted in the 1970s. Arguably however, the potential for good and harm is orders of magnitude larger with gene editing and clearly there is no unified global stance.

Although much of the recent discussion has been within the scientific community and the Academy, the extensive news coverage around the announcement of the birth of the Chinese twins has raised the public's consciousness about gene editing and gene therapy. Over the last few years, attempts have been made to understand how the public views gene editing. In a 2017 report of more than 1000 individuals from 10 European countries and almost 12,000 from the United States respondents were asked to assess the moral acceptability of vignette based decisions. Four scenarios, depicting adult disease therapy, prenatal disease therapy, adult enhancement or prenatal enhancement, were rated. Across the 11 countries, disease treatment is consistently supported far greater than enhancement. Interestingly support across all countries was also greater for adult as compared to prenatal intervention. Uniformly high support was voiced for adult disease therapy whereas the concept of prenatal enhancement showed broad rejection. Adult enhancement and prenatal disease therapy showed very diverse opinions [18]. Support for adult therapeutic interventions has been found in other studies as has the rejection of enhancement interventions. The finding of variable opinions regarding prenatal disease therapy is of interest as that has been considered to be an appropriate interventional goal. The ability to alter the very "code of life" has great ethical implications. As is frequently the case, the ethical discussions and decisions are reactive rather than proactive. Preparing our future physicians to thoughtfully engage in and serve as honest brokers for such ethically and sociologically challenging discussions should be a priority for our medical education institutions.

Without minimizing the ethical issues, moving gene therapy and gene editing into mainstream clinical practice will require successful navigation of multiple additional barriers. Vectors and editing agents will need to be produced in industrial quantities. Decisions must be made whether to pursue a cafeteria of "off-the-shelf" solutions or to focus on individualized approaches. The economics are complex. How does one price a curative therapy? Should the price point be what it would traditionally cost to treat a chronic disease over a lifetime or simply reflect the costs of production and delivery? For many diseases, this will challenge the prevailing business model of pharmaceutical companies. Obviously, it also has implications for clinical facilities and physicians. The potential disruption emanating from gene therapy and gene editing to our existing delivery and economic models remains to be fully revealed. However, it is likely that its influence will be felt to varying degrees throughout the mid-century clinical practice.

Expectations for gene therapy have ranged from a panacea for an array of human maladies to, at best, a niche therapy with disappointingly few applications. Future reality is likely in the middle. It will prove to be a valuable therapeutic modality for a growing number of disease states [19]. More perplexing is the debate about the propriety of gene editing and gene therapy for human enhancement. Whether it be "designer babies," superior intelligence or physical abilities there will likely be clinical systems, if not in the United States, then elsewhere in the world, that will employ these techniques for human enhancement wherever there is a market for such interventions.

Advances in our abilities to determine risk for disease based on genomic analyses and introduce genetic material to compensate for, repair or replace "defective" genes will modify and disrupt the practice of clinical medicine as we now know it. Given this likelihood, medical educators must ensure that future physicians are well-versed in not only the biologic but also the ethical and social implications that accompany the technology.

References

[1] http://omim.org. (Accessed March 21, 2019).
[2] Manrai AK, Ioannidis JPA, Kohane KS. Clinical genomics from pathogenicity claims to quantitative risk estimates. JAMA 2016;315(12):1233–4.

[3] Manrai AJ, Funke BH, Rehm HL, Olesen MS, Maron BA, Szolovits P, et al. Genetic mis-diagnoses and the potential for health disparities. N Engl J Med 2016;375(7):655–65.

[4] Slavin TP, Van Tongeren LR, Behrendt CE, Solomon I, Rybak C, Nehoray B, et al. Prospective study of cancer genetic variants: variation in rate of reclassification by ancestry. J Natl Cancer Inst 2018;110(10):1059–66.

[5] Mersch J, Brown N, Pirzadeh-Miller S, Mundt E, Cox HC, Brown K, et al. Prevalence of variant reclassitication following hereditary cancer genetic testing. JAMA 2018;320(12):1266–74.

[6] Ceyhan-Birsoy O, Murry JB, Machini K, Lebo MS, Yu TW, Fayer S, et al. Interpretation of genomic sequencing results in healthy and ill newborns: results from the BabySeq Project. Am J Hum Genet 2019;104:76–93.

[7] https://www.ncbi.nlm.nih.gov/gtr/. (Accessed March 25, 2019).

[8] Schwartz MLB, Williams MS, Murray MF. Adding protective genetic variants to clinical reporting of genomic screening results: restoring balance. JAMA 2017;317:1527–8.

[9] Abul-Husn NS, Manickam K, Jones LK, Wright EA, Hartzel DN, Gonzaga-Jauregui C, et al. Genetic identification of familial hypercholesterolemia within a single U.S. health care system. Science 2016;354(6319):aaf7000. https://doi.org/10.1126/science.aaf7000.

[10] Kaufmann KB, Buning H, Galy A, Schambach A, Grez M. Gene therapy on the move. EMBO Mol Med 2013;5:1642–61.

[11] Dunbar CE, High KA, Joung JK, Kohn DB, Ozawa K, Sadelain M. Gene therapy comes of age. Science 2018;359(6372):eaan4672. https://doi.org/10.1126/science.aan4672.

[12] ClinicalTrials.gov. (Accessed March 22, 2019).

[13] Komaroff AL. Gene editing using CRISPR why the excitement? JAMA 2017;318(8):699–700.

[14] Kosicki M, Tomberg K, Bradley A. Repair of double-strand breaks induced by CRISPR–Cas9 leads to large deletions and complex rearrangements. Nat Biotechnol 2018;36:765–71.

[15] National Academies of Sciences, Engineering, and Medicine. Human genome editing: science, ethics and governance. Washington, DC: The National Academies Press; 2017.

[16] Scientist claims first gene-editing of human embryos, igniting storm of controversy. Issues Sci Technol 2018;35(2):20.

[17] Lander ES, Bayli F, Zhang F, Charpentier E, Berg P, Bourgain C, et al. Adopt a moratorium on heritable genome editing. Nature 2019;567:165–8.

[18] Gaskell G, Bard I, Allansdottir A, Vieira da Cunha R, Eduard P, Hampel J, et al. Public views on gene editing and its uses. Nat Biotechnol 2017;35(11):1021–3.

[19] Naldini L. Gene therapy returns to center stage. Nature 2015;526:351–60.

Additional resources

- Mukherjee S. The gene an intimate history. New York, NY: Scribner; 2016.
- Genetics Home Reference. Your guide to understanding genetic conditions. NIH National Library of Medicine. https://ghr.nlm.nih.gov/.
- MyGene2. A site dedicated to the sharing of health and genetic information for families with rare diseases and connecting them with clinicians and researchers. https://mygene2.org.
- National Academies of Sciences, Engineering, and Medicine. Human genome editing: science, ethics, and governance. Washington, DC: The National Academies Press; 2017. https://doi.org/10.17226/24623.

CHAPTER

11

Regenerative medicine

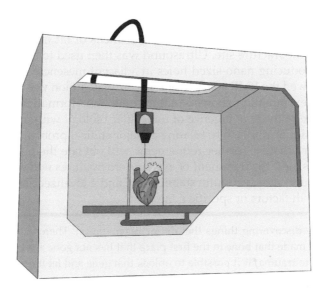

Regenerative medicine is the process of creating living, functional tissues to repair or replace tissue or organ function loss due to age, disease, damage, or congenital defects. [1]

As is implied by this definition, regenerative medicine encompasses numerous approaches and bridges multiple disciplines. Building on the recent dramatic progress in stem cell biology [2], genome editing, elucidation of the role of extracellular matrix in cellular differentiation [3, 4], the roles of signaling pathways in the control of normal development [5], three-dimensional printing and material sciences [6, 7], to name just a few, has resulted in significant progress toward the goal of regenerating or replacing diseased or damaged human tissue and organs [8]. Although there are few FDA approved regenerative medicine therapies, the field is rapidly evolving with potentially exciting developments in many areas [9]. Let us consider a few examples.

Implementing Biomedical Innovations into Health, Education, and Practice
https://doi.org/10.1016/B978-0-12-819620-5.00011-4

It is estimated that about 100,000 people yearly suffer from a nonunion bone fracture in the United States. If there is bone missing or it is badly splintered, autologous bone grafting is often required. How to enhance bone healing without the need for a graft has been a focus of investigation for years. The approach most studies have taken involved filling bone gaps with collagen and then introducing bone morphogenetic proteins (BMP), or using a vector to deliver an osteogenic BMP gene, to induce mesenchymal stem cell differentiation into bone producing cells—osteocytes. Unfortunately, results have been mixed and overall disappointing. In an innovative study, following the standard practice of using a collagen matrix to fill the bone gap and waiting for mesenchymal stem cells to infiltrate the scaffold material, a solution containing the BMP gene and gas-filled micron size bubbles encased in shells of fat molecules was injected into the fracture site. Ultrasound was then used to burst the microbubbles producing nano-sized holes in adjacent mesenchymal stem cells which allowed the BMP genes to enter. Complete union was obtained and histologic examination showed mature new bone formation [10]. This is an example of the convergence of molecular biology with engineering to create a unique approach to treating a major clinical problem. It allows us to consider whether further refinements will obviate the need for future bone grafts, or if the treatment of even severe fractures will, in the not too distant future, simply require stabilization and a localized injection delivering growth factors or specific genes?

> "We're discovering things that define regeneration. There is a part of the gene that made that bone in the first place that has not gone anywhere, … (In response to trauma) is it possible to unlock that gene and let it regenerate that bone? I don't think that's far-fetched."
>
> *Edward Schulak*

Sensorineural hearing loss is a common condition affecting millions worldwide. Unlike the natural hair cell regeneration found in birds, amphibians, and fish, human beings have no ability to replace lost hair cells in the cochlea. An intense focus of research has been on the processes that result in hair cell regeneration and how it can be replicated in mammals. One of the challenges in humans is that cochlear stem cells essentially disappear within a few weeks of life. However, investigators have shown that stem cells derived from bone marrow and induced pluripotent stem cells (iPS) can be made to differentiate into hair cells in vitro. In adult guinea pig models of deafness, hearing improvement and the appearance of new hair cells has been achieved in vivo [11]. Will advances in stem cell biology allow us to reverse hearing loss and regenerate cochlear hair cells? Will there be a "cure" through regeneration of hair cells for deafness?

While not generally thought of as "organ system failure" or disease in the classic sense, the development of sarcopenia and frailty in the elderly due to changes in muscle mass and strength leads to increasing difficulty in activities of daily living, mobility and eventually loss of independence. Multiple interventions to reverse this decline and improve muscle function targeting different aspects of sarcopenia have shown limited or no efficacy. While still early in our understanding, if research identifies means to prevent or halt sarcopenia it has major implications given the aging population in the United States and around the world [12].

Transplantation is a well-established intervention for a multitude of failing organs. However the chronic shortage of suitable organs, the logistics of organ harvesting and transplant and long-term issues with immunosuppression and rejection has led to the ongoing quest to develop ex vivo organs for implantation or develop effective in vivo therapy for damaged tissues to restore function.

Xenotransplantation, implanting in humans organs grown in other animals—most notably pigs, has been proposed for decades. However violent immune responses curtailed progress. Recently genetic modifications of the pigs have resulted in organs that are well tolerated when transplanted into primates.

However many see xenotransplantation as a "stopgap" measure. The goal is to develop artificial organs populated with the individual's own cells or closely matched commercially available cell populations, eliminating or greatly reducing the need for immunosuppression. The complexity of replacing whole organs is enormous. Appropriate materials to form the cellular scaffolding must be developed. Manufacturing techniques must be devised to accurately position different types of cells within the requisite 3D structure. Growth factors must be delivered precisely to ensure differentiation of implanted stem cells into the tissues that make up the organ and its vascular supply. These are only some of the hurdles. While recognizing the enormous number of issues that still must be resolved, the rapid advances in bioprinting, allowing precise buildup of cellular scaffolding, cells and biochemicals, holds promise for the fabrication of a wide range of tissues and organs [13]. An intermediary step is the development of intestinal, brain, kidney, heart and other tissue organoids, providing additional tools for exploring the complex network of interactions required for differentiation of stem cells into functional organs.

Rather than developing an artificial organ for implantation, a parallel strategy is the induction of an individual's endogenous stem cells to dedifferentiate and repair damaged tissue. Reprogramming pancreatic exocrine cells into beta-like cells and cardiac fibroblasts into functioning myocytes, are examples of ongoing work using transcription factors to reprogram cells in vivo. Possibly related is the intriguing observation that under optimal conditions cells can recover from apoptosis and other forms of programmed cell death. While not universally embraced by the research community, it appears that cells have the capacity to recover normal morphology and function following an insult that had been thought to irreversibly lead to cell death. This observation has led to speculation that understanding the multitude of processes involved in apoptosis might lead to interventions to minimize or reverse the damage from myocardial ischemia, cerebrovascular accidents, spinal cord injury, neurodegenerative processes and trauma or diseases that affect any nonrenewable cells [14, 15]. Will it be possible to reverse cell senescence and death thereby increasing longevity and prolonging functional life? Will we be able to repair damaged tissue in situ?

Acknowledging that remarkable progress has been made in regenerative medicine, there is real danger that the hype has raised the public's expectations far beyond the reality we can deliver. Hopes have been raised that arthritic conditions can be "cured" with in vivo cartilage repair, that low output cardiac conditions due to myocardial damage can be reversed, that a host of neurologic conditions ranging from amyotrophic lateral sclerosis (ALS) to paralysis can be a treated, that organ shortages for transplantation will be ameliorated, and the list goes on. Not surprisingly, given the widespread attention and publicity that the potential of stem cell therapy and regenerative medicine has generated, unscrupulous "clinics" marketing directly to consumers have opened their doors, promoting unsupported therapeutic uses. While governmental authorities have been successful in some countries to curtail or close such clinics, the problem remains [16]. This highlights the importance of trustworthy information sources and the need for scientifically rigorous studies to evaluate potential clinical applications. It also is a cautionary note that premature claims emanating from basic science research findings that have a potential clinical application for serious diseases may inadvertently result in increased patient suffering.

Of all the technologies considered as potential midcentury clinical practice disruptors, the impact of regenerative medicine remains the least defined. That said, it arguably has the greatest potential for disrupting clinical medicine as we now practice it. Consider for a moment, if the envisioned potential applications to chronic problems of aging such as osteoarthritis and joint disorders materialize, it could have a positive impact on millions and markedly reduce the need for orthopedists to implant joint prostheses. If nerve regeneration is realized, being able to reverse the paralysis resulting from spinal cord injury would dramatically change the injured patient's prognosis and minimize the need for assistive care. If organs for transplant are readily available, it would decrease or eliminate the need for renal dialysis and the associated clinical services that nephrologists provide. If complex fractures and severe wounds were treated with injections of growth factors and genes the decreased morbidity and enhanced quality of patients' lives would be considerable.

Time will tell whether the promise of regenerative medicine becomes a widespread clinical reality. I am confident that many aspects of regenerative medicine will become part of clinical practice. Hence, we need to prepare our medical students and residents to be sophisticated consumers of the claims for therapeutic benefit already promulgated in this growing field. Our future physicians must be trained to recognize what is and is not reasonable, viable, and realistic for their patients. This requires not only a deep understanding of molecular biology but also material sciences and related engineering fields.

References

[1] Regenerative medicine. National Institutes of Health Fact Sheet; 2010. https://report. nih.gov/nihfactsheets/ViewFactSheet.aspx?csid=62.

[2] Tewary M, Shakiba N, Zandstra PW. Stem cell bioengineering: building from stem cell biology. Nat Rev Genet 2018;19(10):595–614.

[3] Grande DA. Important milestones on the way to clinical translation. Nat Rev Rheumatol 2017;13(2):67–8.

[4] Hussey GS, Dziki JL, Badylak SF. Extracellular matrix-based materials for regenerative medicine. Nat Rev Mater 2018;3:159–73.

[5] Tabar V, Studer L. Pluripotent stem cells in regenerative medicine: challenges and recent progress. Nat Rev Genet 2014;15:82–92.

[6] Madl CM, Heilshorn SC, Blau HM. Bioengineering strategies to accelerate stem cell therapeutics. Nature 2018;557:335–42.

[7] Kang HW, Lee SJ, Ko IK, Kengla C, Yoo JJ, Atala A. A 3D bioprinting system to produce human scale tissue constructs with structural integrity. Nat Biotechnol 2016;34(3):312–9. https://doi.org/10.1038/nbt.341.

[8] Takata N, Eiraku M. Stem cells and genome editing: approaches to tissue regeneration and regenerative medicine. J Hum Genet 2018;63:165–78. https://doi.org/10.1038/ s10038-017-0348-0.

[9] Mao AS, Mooney DJ. Regenerative medicine: current therapies and future directions. PNAS 2015;112:14452–9.

[10] Bez M, Sheyn D, Tawackoli W, Avalos P, Shapiro G, Giaconi JC, et al. In situ bone tissue engineering via ultrasound-mediated gene delivery to endogenous progenitor cells in mini-pigs. Sci Transl Med 2017;9(390):eaal3128. https://doi.org/10.1126/scitranslmed. aal3128.

[11] Diensthuber M, Stover T. Strategies for a regenerative therapy of hearing loss. HNO 2018;66(Suppl 1):S39–46.

[12] Miller RR, Roubenoff R. Emerging interventions for elderly patients—the promise of regenerative medicine. Clin Pharmacol Ther 2019;105(1):53–60.

[13] Park KM, Shin YM, Kim K, Shin H. Tissue engineering and regenerative medicine 2017: a year in review. Tissue Eng Part B Rev 2018;24(5):327–44.

[14] Gong YN, Crawford JC, Heckmann BL, Green DR. To the edge of cell death and back. FEBS J 2019;286:430–40.

[15] Choi CQ. Back from the brink of death. Apoptosis and other types of programmed cell death appear to be reversible. Scientist 2019;2:32–9.

[16] Sipp D, Caulfield T, Kaye J, Barfoot J, Blackburn C, Chan S, et al. Marketing of unproven stem cell based interventions: a call to action. Sci Transl Med 2017;9:eaag0426. https:// doi.org/10.1126/scitranslmed.aag0426.

Additional resources

- Institute of Medicine and National Research Council. Stem cell therapies: opportunities for ensuring the quality and safety of clinical offerings: summary of a joint workshop by the Institute of Medicine, the National Academy of Sciences, and the International Society for Stem Cell Research. Washington, DC: The National Academies Press; 2014. https://doi.org/10.17226/18746.
- National Academies of Sciences, Engineering, and Medicine. Mitochondrial replacement techniques: ethical, social, and policy considerations. Washington, DC: The National Academies Press; 2016. https://doi.org/10.17226/21871.
- Blau HM, Daley GQ. Stem cells in the treatment of disease. N Engl J Med 2019;380(18):1748–60.
- Haddock R, Lin-Gibson S, Lumelsky N, McFarland R, Roy K, Saha K, et al. Manufacturing cell therapies: the paradigm shift in health care of this century. NAM Perspectives. [Discussion paper]. Washington, DC: National Academy of Medicine; 2017. https://doi.org/10.31478/201706c. https://nam.edu/manufacturing-cell-therapies-the-paradigm-shift-in-health-care-of-this-century/.
- National Academies of Sciences, Engineering, and Medicine. Navigating the manufacturing process and ensuring the quality of regenerative medicine therapies: proceedings of a workshop. Washington, DC: The National Academies Press; 2017. https://doi.org/10.17226/24913.
- National Academies of Sciences, Engineering, and Medicine. Exploring the state of the science in the field of regenerative medicine: challenges of and opportunities for cellular therapies: proceedings of a workshop. Washington, DC: The National Academies Press; 2017. https://doi.org/10.17226/24671.

CHAPTER

12

Precision medicine

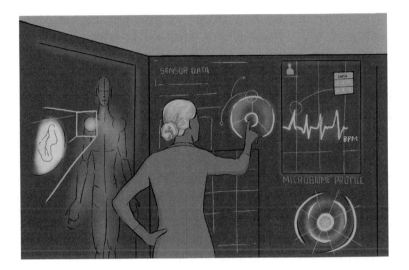

Background

Diagnosis is the foundation of medicine. Accurately and precisely defining a patient's condition does not assure effective treatment, but it is unequivocally the place to start. [1]

Advances in the technology that enable scientific discovery have resulted in inflection points in clinical medicine. Earlier in this book, we discussed how the microscope led to discoveries that have shaped much of the current practice of medicine. These types of discoveries resulted in fundamental changes in our characterization of disease and in the mental models physicians developed for diagnosis and management of specific conditions. Ultimately, these changes were reflected in the way future physicians were trained.

Implementing Biomedical Innovations into Health, Education, and Practice
https://doi.org/10.1016/B978-0-12-819620-5.00012-6

Now, the convergence of computational, technologic, and scientific advances is enabling another inflection point—"precision medicine." For much of the last century, infectious diseases were the only maladies whereby the diagnosis entailed identification of a specific etiology and therapy was aimed specifically at that causative agent, namely the involved pathogen. This is now changing due to advances in molecular biology. Disease entities, such as cancers, that for decades were diagnosed based upon their histology and categorized by the organ in which they originated, may actually be due to different molecular mechanisms despite originating in the same tissue. Likewise cancers originating in different organs and with very different histologic features may share the same pathophysiologic mechanism. Yet, cancers continue to be named and often treated based upon their histology rather than the molecular mechanism.

Advances in sequencing and computational analyses are facilitating the discovery of disease related genetic alterations across a broad spectrum of maladies. Thousands of associations have been reported of single and multiple nucleotide polymorphisms for genetic abnormalities specific to diagnoses ranging from autism to schizophrenia to macular degeneration. Studies are beginning to unravel the complex interactions between our microbiome and health and disease. This is the context in which the genesis of the idea and the vision embodied in the term "precision medicine" arose.

Motivated by the realization that there was a need for a disease taxonomy that reflected current knowledge, the National Academies of Science, Engineering and Medicine charged the Committee on a Framework for Developing a New Taxonomy of Disease to "explore the feasibility and need, and develop a potential framework, for creating a 'New Taxonomy' of human diseases based on molecular biology" [2].

The 2011 monograph, *Toward Precision Medicine: Building a Knowledge Network for Biomedical Research and a New Taxonomy of Disease* [3], which arose from the Committee's deliberations, endorsed the term "precision medicine." The Committee's recommendations were expansive and set forth a vision that would unite biomedical researchers and public health and clinical medicine practitioners around the common goal of enhancing understanding and treatment of disease. This is a long-term goal realized through development of massive databases that would include patients' genomes, microbiomes, exposomes (a subsection of the environment), epigenomes, physiologic data, signs and symptoms, and other relevant information. The data would be analyzed using sophisticated machine learning tools, resulting in a deep understanding of the pathways and pathophysiology of diseases. Further, this conceptualization of health and disease provides a unique opportunity to catalyze additional targeted research and to facilitate the expansion of clinicians' abilities to tailor interventions to the unique characteristics of their patients.

Let us consider the magnitude of what is being proposed. Considering only the human genome (and not the microbiome that is estimated to be 100 times larger, or any of the other variables such as the proteome, transcriptome, and exposome) the challenge of making actionable interpretations of this massive amount of data is monumental. The human genome includes approximately 20,000 protein coding genes [4]. There are approximately 9 billion possible single nucleotide variants. This does not include copy number variants and insertions and deletions. Given that each variant can be found in either the homozygous or heterozygous state, the number of variant combinations is essentially infinite. Over 4.5 million missense variants have already been found in the Genome Aggregation Database. Nearly all of these are rare and only 2% have any sort of clinical interpretation, over half of which are of uncertain significance [5].

Thousands of genomic variants have been reported to be associated with Mendelian inherited disease, but the evidence of disease association is variable. This variability in the level of evidence used to implicate a connection between a gene and a disease has resulted in erroneous associations that are now embedded in the literature and are all too often included in reports to clinicians. As a consequence, a great deal of work is now ongoing to ascertain the validity of gene—disease associations [6]. One approach is to develop and aggregate ever larger data sets of genome wide association studies (GWAS) including individuals from multiple ancestry population groups to determine whether apparent associations are simply due to biased sampling or if indeed there is a causative link. In addition, longitudinal studies of "normal" people with identified pathogenic or potentially pathogenic genomic variants are needed to see whether they ultimately manifest the disease. And most importantly, research needs to be done to determine the mechanism whereby the genomic alterations lead to disease.

"Not all (genetic) variants are created equal. We know that 99.999% of variants have absolutely no effect and there's ways to prove this. We can do this genetically, we can look at issues of selection both positive and negative selection of variants, we can do allele frequency in the population. I think this is some of the fundamental training that's missing. (in medical education).

How do you decide that a variant is pathogenic or if it's benign? .... If the clinician had a little bit of knowledge there are ways to check this pretty quickly and you can say, 'Oh, this is likely going to be benign.'

There is software to help guide them, but I think what is missing is this fundamental understanding about what population genetics tell us."

Evan Eichler

But developing the robust human genome disease association data-bases is only one facet of the problem. An understanding of the relation-ships between our microbiome and health and disease and the databases necessary to characterize those relationships is an even bigger challenge. And then add in physiologic variables, biomarkers, geospatial and envi-ronmental data, and behavioral information and the magnitude of the task encompassed by the term "precision medicine" emerges. Even with the most sophisticated machine learning programs, the computational chal-lenges to analyze the resulting databases are monumental. The sobering reality is that the journey to precision medicine's ultimate goal, tailored therapy at the individual level, will require an incredible amount of work, time, and training.

But progress is being made on several fronts. Let us consider some examples.

Cancer

Cancer researchers and clinicians are familiar with the impact of ge-nomic alterations on risk. For example, BRCA 1/2 mutations have been shown to have a strong correlation with increased risk for breast and ovar-ian cancers [7]. Clinical experience suggests that heritable cancers are not rare. However known genetic variants account for only a limited number of apparent familial cases. Data analysis using The Cancer Genome Atlas (TCGA) of over 10,000 patients with 33 different types of cancer revealed 853 pathogenic or likely pathogenic germline variants in 8% of the cases. Additionally, shared alterations were found across cancer types [8]. While this study greatly expanded the number of identified heritable genomic variants in cancer patients, the sample size would need to increase dra-matically (estimate at least 100,000) in order to approach an 80% likeli-hood of identifying rare predisposing gene variants with 95% penetrance. Identification of potential inherited mutations that increase the risk for cancer is only a first step, albeit an important one. To maximize the poten-tial of precision medicine for prevention and/or cure, the mechanism(s) by which a mutation initiates disease development and the pathways in-volved must be worked out.

Acute lymphoblastic leukemia (ALL), the most common form of leu-kemia in children, is an example where progress is being made in under-standing disease initiation. ALL evolves in two discrete steps. The first step occurs in utero with the generation of a covert pre-leukemic clone by fusion gene formation or hyperdiploidy. Genomic analysis can iden-tify the presence of this predisposing genetic mutation in newborns. In approximately 1% of children with the mutation, postnatal secondary genetic changes drive conversion to overt leukemia. Studies suggest that microbial exposures early in life are protective but later infections may trigger the critical secondary mutations [9]. The intriguing question remains as to whether it may be possible to greatly reduce or eliminate ALL by identifying the predisposed infants and ensuring early in life that their immune systems are challenged by a host of microbes.

Clinicians have long known that patients' responses to chemotherapeutic agents differ. Guidelines have been developed to provide clinicians a "rational" approach to pharmacologic therapy. However, the guidelines are based on large populations of patients with a given cancer diagnosis and a therapeutic trial is required to determine if a patient will respond as desired. As our understanding of the pathophysiologic mechanisms responsible for malignancies increases, the use of genomic sequencing and biomarkers to identify cancers with actionable pathways, rather than relying on histologic based diagnoses, will grow.

An example is fusions involving one of three tropomyosin receptor kinases (TRK) that occur in a variety of pediatric and adult cancers. In a study of a highly selective TRK inhibitor, patients, age of 4 months to 76 years, with TRK fusion positive cancers were enrolled. Seventeen unique cancer diagnoses were included. Independent of age or histologic tumor type, the overall treatment response rate was 75% [10]. Findings such as this reinforce the importance of "precision cancer therapy" which is driven by the pathophysiologic mechanism and is agnostic to the tissue of origin.

To add to the complexity, recent observations of heterogeneous response to immune checkpoint inhibitors in melanoma have implicated the microbiome. In a study of over 100 patients with metastatic melanoma receiving anti-programmed cell death 1 protein (PD-1) immunotherapy responders were found to have significantly more diversity and a different composition of their gut microbiome as compared to nonresponders [11]. A parallel observation is that recent antibiotic use, known to alter the gut microbiome, appears to be associated with decreased responsiveness to chemotherapy. Is there a causal link between the gut microbiome and chemotherapeutic efficacy or simply a correlation? Recognizing the important role that the gut microbiota plays in immune function, these observations are noteworthy. However, the mechanism of this modulation, if indeed it is causal, remains to be elucidated.

The promise of chemotherapeutic decisions based on the specific characteristics of a patient's tumor is just beginning to be realized. This has resulted in intensive research seeking to identify genetic alterations and biomarkers that can be utilized to prescribe chemotherapy that is efficacious against the malignancy's pathway. While notable successes have been observed, much remains to be done. Future physicians and biomedical researchers will need to understand the rate of somatic mutations, the patient's microbiome, and subsequent pathologic pathway evolution within a tumor that leads to "resistance."

In addition to using genetic analyses to determine the sensitivity of a tumor to chemotherapy, side effects from the chemotherapy on the patient depend upon the individual's genome. Vincristine, a frequently used chemotherapeutic agent, can be associated with a major adverse side effect of peripheral neuropathy which can be sensory or motor. Historical observations, including a decreased frequency in patients of African ancestry, suggested that there might be a genetic determinant to vincristine associated neuropathy. Now studies have implicated *CEP72* promoter polymorphism with the neuropathy [12].

As more evidence accumulates regarding how best to tailor chemotherapy based upon the characteristics of an individual patient's tumor and their genetically determined tolerance for potential side effects, the educational challenge is insuring that our future clinicians are equipped to apply the principles and knowledge in the care of their patients—to practice "precision oncology."

Pharmacogenomics

Advances in our understanding of how an individual's genetics affects the efficacy of pharmacologic therapies are not limited to oncology. As research on the genetics related to an individual's drug metabolism increases, the value of genomic screening to ascertain efficacy also increases. This marriage of pharmacology and genomics is termed pharmacogenomics. A wide range of pharmacologic agents ranging from anticoagulants [13], to antidepressants [14], to smoking cessation medications [15] have been studied. While still in its infancy, pharmacogenomics holds broad promise for tailored, patient specific therapy. Knowing that an individual will not benefit from a specific pharmacological intervention and could even potentially be harmed by it not only avoids the financial waste but also the morbidity that might result from delayed or ineffectual treatment.

While it makes sense intuitively to use genetic testing in order to guide pharmacologic therapy, the clinical reality is that even well studied common polymorphisms account for only a fraction of observed drug effects. Moreover, the marketing of clinical genetic tests to guide pharmacologic therapy does not necessarily indicate robust evidence of improved patient outcomes. Despite the current paucity of instances where genetic determination is required for therapeutic decision-making commercial laboratories such as Intermountain's RxMatch—Pharmacogenomics [16] provide testing services so "physicians can fast-track finding the right medication and dose for patients." Using the DNA sample submitted by the patient, the laboratory performs a series of analyses and sends a report to the requesting physician. The report can be used to guide drug choices and appropriate dosing for a wide range of medications including opioids, statins, immunosuppressants and antidepressants. While the promise of pharmacogenomics is great, the current reality is limited. However, I expect that by mid-century, rational pharmacologic therapy based on an individual's genetics, and possibly their microbiome, will be well-established for a wide array of diseases.

Behavior and environment

Oftentimes lost in the discussions about precision medicine is the totality of factors that must be considered as they affect an individual's health and disease. As technologic advances have allowed precise genetic determinations, the majority of "precision medicine" studies have focused on the genome. But people have linked behaviors to health and disease for centuries. Obvious factors such as risk-taking behaviors, drinking and driving, smoking, and inactivity have been linked to undesirable outcomes and resulted in efforts to change those behaviors across society.

Although behaviors are usually thought to be independent of inherited influence, there have been intriguing findings that there are genomic predispositions to some behaviors that have important health and disease implications. A recent study of smoking reported dozens of single nucleotide variants that were associated with smoking behavior [17]. But causal mechanisms, if any, remain to be elucidated. The importance of such studies is that they begin to highlight the complicated interrelationships among the various determinants of health and disease. They also underscore the major gaps in our knowledge base regarding these interactions. We are only beginning to develop an understanding of the effects of behaviors on the genome through epigenetic changes, the microbiome, genomic expression, and a multitude of physiologic processes. The importance of this work is that it allows us, or some would say forces us, to see individuals not as a collection of isolated "risk factors" but rather as an integrated whole.

In addition, an individual's environment plays a major role in their health and the development of disease. This includes the built environment, pollutants and other environmental exposures, their social environment and the whole array of social determinants of health that are inexorably linked to their behaviors and their immediate environment.

As was recently stated in a National Academy of Medicine perspective, "Health is the product of our experiences layered onto the biological matrices we inherit. Those experiences begin at conception, and, through the intersecting influences of genetics, environment, social circumstances, behaviors, and medical care, health emerges and takes. Each of us represents, in essence, a complex system in constant and dynamic interface with other systems that shape our fates in manners great and small" [18].

Much has been written about the importance of social determinants of health and for many outcomes they are more important than medical care. Variables including education, income, housing, nutrition, and occupation are highly correlated one with the other and health outcomes such as life expectancy. But untangling the web of correlation and causation is vexing. And likely it differs when considering inner-city urban versus rural populations, different ethnic groups, first-generation immigrants, and the like.

There are direct and indirect environmental effects on the individual's health and well-being. Lead exposure and other pollutants can result in an array of diseases ranging from neurological and cognitive impairment to skin irritation. Abusive physical and psychological relationships can lead to depression, self-harm, and poor mental health. Food insecurity, substandard housing, and inadequate educational opportunities—often found in low social economic strata individuals—may result in malnutrition, altered cognitive development, asthma, and an array of mental health disorders.

The relationship between environmental chemical exposures such as polychlorinated biphenyls (PCBs), dioxins, and bisphenol A (BPA) and disrupted endocrine functions are beginning to be understood. But there likely remains a large number of previously unrecognized environmental factors that may directly lead to disease, result in epigenetic changes, impact the microbiota, negatively affect individual's health and even result in abnormalities that persist to future generations.

The likelihood of one's patients being exposed to environmental hazards can be estimated but requires accessing and understanding public databases rarely utilized in medical settings. Resources such as geospatial mapping of toxic waste sites and contaminated groundwater plumes, maps of airborne dispersion of particulates and pollution from power plants and industrial sites, and census track data on housing age and the probability of lead exposure can be useful. The full extent of environmental impacts on our health and the development of disease remains to be elucidated. As our future physicians embark upon their clinical careers, the relevance of such disparate databases to their clinical practice should be as apparent as traditional biochemical blood tests.

In order for the power of precision medicine to be realized, it is necessary that all factors affecting an individual's health and the development of disease be considered. Indeed, behavior and environment are likely far more important than genetics for many diseases. The imprecision of our tools is an impediment toward progress in fully understanding the contributions of behavior and environment. However, a greater impediment is the limitation imposed by the reductionist model of understanding. Precision medicine requires that we embrace a comprehensive model of health and disease which inevitably leads to a complex web of interactions rather than a simple cause and effect. This requires a fundamental change in mindset.

"There is a tension in precision medicine right now. Typically you hear people talk about precision medicine defining subcategories of disease and other people will say, "No, precision medicine is only going to drive costs because there will be too many subcategorizations."

The people that are on the cost driver-side are just thinking of these endless subpopulations. The people that are thinking about it as a positive thing are recognizing that, now wait a minute, we get expensive drugs to 10 people and only one person needs it. You know it can be a cost savings.

I think we're far away from understanding individual uniqueness well …. It also misses the environmental influence. If you look at Barbers in North Africa they're genetically fairly homogeneous. But they can be agrarian, they can be urban, or they can be coastal. If you look at those three different populations that are genetically pretty similar but are environmentally distinctive, they have completely different disease burdens."

James Madara

What does this mean for future physicians?

Consumer genomic testing

While the research necessary to understand the implications of genomic variants for disease predisposition, prognosis and therapy is ongoing, the marketing of genetic analysis to the public is widespread. Technologic advances mean that for under $200 anyone can obtain their genetic information. Simply spit in a tube, send it off, and a few weeks later you will receive a report that will include information on everything from earwax composition to eye color to the potential for some diseases with established genetic links.

When an individual receives unexpected results such as two copies of the APOE4 gene (associated with approximately a 10-fold increased risk for Alzheimer's) or harmful BRCA 1/2 variant (about 70% chance of developing breast cancer by age 80 as well as increased risk for ovarian and other cancers), the task of making sense of this information increasingly falls on the clinician. While genetic counselors are well-positioned to work with parents or patients who are dealing with established genetic conditions, their numbers and perhaps also their training, are insufficient to educate the tens of thousands of individuals who are opting for commercially available genetic analyses. This will require that clinicians not only possess the necessary knowledge base but also the appropriate communication skills to explain the results. The complexities of multi-genomic diseases and the fact that having a genetic variant that puts one at risk for disease does not necessarily mean that the disease will occur are challenging concepts to convey.

The anxiety experienced by individuals receiving a report of previously unrecognized genomic risk is well recognized. Less appreciated is that some individuals will interpret a report indicating absence of known genomic risk as meaning that they will not get the disease in question and therefore not adhere to prudent recommendations. Both are realities that clinicians must be prepared to address.

A recent survey of primary care physicians showed that less than 15% felt confident interpreting genetic tests. Further confounding the problem is the fluid state of our understanding of the significance attributable to different mutations (permanent changes in the nucleotide sequence), polymorphisms (variants with a frequency > 1%) and combinations thereof, and the difficulty that even experts have in interpreting the myriad of sequence variants [19].

Diagnostic advances

The power of genomic analyses to aid in the specific diagnosis of disease has been a major, albeit limited, advancement. But, genomic analysis is only one of an entire suite of tools that will increasingly be available to clinicians. The development and availability of laboratory studies to characterize the transcriptome, metabolome, and proteome will provide additional potentially clinically relevant biomarkers for the diagnosis of disease and for monitoring its progression.

A major challenge for clinicians will be the ability to determine what is "normal," what the threshold for diagnosis is, and when to intervene. We know that asymptomatic healthy individuals can harbor pathologic microorganisms and that no therapy is indicated. We are just beginning to explore what it means to have genomic variants or biomarkers in apparently normal disease-free individuals.

An example of the conundrum physicians will increasingly encounter is a study seeking to identify endometrial cancer at an early stage. One hundred seven women had uterine lavage prior to undergoing hysteroscopy and curettage. The lavage fluid was then studied to detect somatic mutations associated with endometrial cancer and compared to the histopathologic analysis. Seven patients diagnosed with endometrial cancer had significant cancer associated gene mutations. Six of the seven patients were stage IA and the other cancer was only detected as a microscopic focus within a polyp. However 51 patients without histopathologic evidence of cancer had cancer associated mutations ranging from 1% to 30% [20]. What do findings such as this mean? Is this the genetic equivalent to the well-known phenomenon of increasing prevalence of histopathologic changes consistent with prostate cancer in men as they age? Or is this an opportunity to intervene at the earliest stage of endometrial cancer development?

Diagnosing a person as having a disease has great implications. In some situations it may mean there is an opportunity for cure. In others it may mean that an individual undergoes needless interventions with the attendant morbidity. Our understanding of the implications for a given individual of the results generated from our increasingly more sophisticated laboratory studies has not kept pace with the technology; and this is just for relatively straightforward observations. As more complex findings of multiple polymorphisms coupled with microbiome data and a multitude of biomarkers become available, the task of the physician to correctly "diagnose" grows exponentially. Clinicians will be faced with such questions and will need to determine whether intervention is required or whether the finding is of little or no clinical significance—often with limited data on which to base their decisions. The physicians of the future will need to be trained not just to understand and interpret the results but also to know what to do with them.

Therapeutic recommendations

The advantages of tailoring therapy to the specifics of the individual, or N of 1 therapy, are undeniable. For this to become a reality, significant advances must be made in our biomedical knowledge as applied to clinical medicine and the whole approach to therapeutic interventions must change. Patients will not be treated based on their cohort status but rather as individuals. While some recommendations such as smoking cessation and avoidance of second-hand smoke will remain for the population as a whole, most other clinical recommendations, including disease surveillance, will be individualized.

Taken to its extreme, every individual will be unique. The implications of the shift from population-based to individualized decision-making for therapeutic interventions are multifold. As recognition of the heterogeneous pathophysiologic basis of many diseases has grown, clinical trials have sought to identify homogenous subgroups of patients for targeted therapy. As such, many diseases, even common ones, are "rare diseases" when defined evermore precisely. When assessing the efficacy of new agents, the standards of evidence are impacted as the ability to conduct randomized controlled trials will be limited. And very importantly, when genomic analyses and readily measurable biomarkers are the basis for determining homogenous but ever smaller subgroups, the numbers of individuals needed to determine a new therapeutic intervention's efficacy may not be achievable.

Full implementation of N of 1 therapy requires that interventions ranging from pharmacologic agents to diet to activity be based on an understanding of the physiologic and pathophysiologic pathways involved. Obviously, this requires knowing the effects of multiple interventions on a specific pathway. While there has been significant progress in the development of targeted pharmacologic agents, relatively little is known about dietary and activity-based "prescriptions." One example is the role of the gut microbiota. It has been implicated in conditions as diverse as obesity, diabetes, coronary artery disease, and neurologic disorders. However a coherent approach to altering the gut microbiota to enhance health and reverse or minimize disease does not exist. Developing an understanding of the influence of diet on the gut microbiota ecosystem will be necessary for appropriate "dietary prescriptions" tailored to the individual's needs. While "lose weight" or "exercise more" may be generally appropriate advice, it is not consistent with the tenants embodied in precision medicine. The physician must be able to fully understand the unique health and disease attributes of each individual and to provide input, treatment and therapy accordingly.

In addition to the necessary expansion of knowledge of therapeutic options, the physician must also know when intervention is indicated along the continuum from high risk to symptomatic disease. Understanding when sufficient information has been obtained to meet threshold criteria for preventive or therapeutic interventions will also be an important role for physicians. While the current reality does not begin to approach this vision, as our scientific knowledge base expands to begin to fully realize the potential of precision medicine and enables tailoring of interventions to the individual for the majority of clinical situations, N of 1 clinical care will radically alter the practice of medicine and the experience of the patient.

"...the dominant theme I think in the future is going to be scientific wellness. That means for the most part that physicians are going to be following people that are relatively healthy. So, their major functions are going to be to enable them to stay healthy and even get healthier, and there are straightforward ways of doing that.

I think the idea of coaching and the idea of these dense personal data clouds all are going to be critical elements, but the really interesting challenge for them (physicians) as they follow these people is that you will over time see transitions from wellness to disease for every single common disease.

What we will develop over the next 10 or 15 years are biomarkers that can pinpoint precisely when the earliest transition for pre-diabetes, or whatever you want, takes place and we'll use the systems thinking to understand what drugs can reverse the disease before it ever manifests itself as phenotype. It is the preventive medicine of the 21st century. You're going to reverse people back to wellness before they ever know they have a disease.

Physicians, with the right computational aides, will realize that this P4 medicine has enormously enhanced what they can do. It's enormously increased their efficiency in keeping people well, and it's enormously enabled them to see the earliest disease transitions and move toward reversal."

Leroy Hood

Precision medicine is the culmination of the advances being made in biomedical science, technologic and computational advances. Importantly, it is a comprehensive vision, including but not restricted to genomic assessments. It will disrupt the practice of medicine as it applies not just to the treatment of disease states, but also to the identification of individuals in the presymptomatic or at risk for disease state. Although initially it will be most impactful in the treatment of diseases such as cancer, progress will also be made to prevent the development of symptomatic disease and restore to normal. Some would characterize this as an audacious goal that realistically will take decades to realize. I see it as a bold but rational next step in how we conceptualize health and disease. To realize this vision requires not only continued scientific and technologic advances, but also a change in the educational focus and preparation of our future physicians.

References

[1] National Research Council. Toward precision medicine: building a knowledge network for biomedical research and a new taxonomy of disease. Washington, DC: The National Academies Press; 2011. p. 79. https://doi.org/10.17226/13284.

[2] National Research Council. Toward precision medicine: building a knowledge network for biomedical research and a new taxonomy of disease. Washington, DC: The National Academies Press; 2011. p. 76. https://doi.org/10.17226/13284.

[3] National Research Council. Toward precision medicine: building a knowledge network for biomedical research and a new taxonomy of disease. Washington, DC: The National Academies Press; 2011. https://doi.org/10.17226/13284.

[4] OMIM. http://omim.org/.

[5] Starita LM, Ahituv N, Dunham MJ, Kitzman JO, Roth FP, Seelig G, et al. Variant interpretation: functional assays to the rescue. Am J Hum Genet 2017;101:315–25.

[6] Strande NT, Riggs ER, Buchanan AH, Ceyhan-Birsoy O, DiStefano M, Dwight SS, et al. Evaluating the clinical validity of gene-disease associations: an evidence-based framework developed by the clinical genome resource. Am J Hum Genet 2017;100:895–906.

[7] Kuchenbaecker KB, Hopper JL, Barnes DR, Phillips KA, Mooij TM, Roos-Blom M, et al. Risks of breast, ovarian, and contralateral breast cancer for BRCA1 and BRCA2 mutation carriers. JAMA 2017;317(23):2402–16.

[8] Huang K, Mashl RJ, Wu Y, Ritter DI, Wang J, Oh C, et al. Pathogenic germline variants in 10,389 adult cancers. Cell 2018;173:355–70.

[9] Greaves M. A causal mechanism for childhood acute lymphoblastic leukaemia. Nat Rev Cancer 2018;18:471–84.

[10] Drilon A, Laetsch TW, Kummar S, DuBois SG, Lassen UN, Demetri GD, et al. Efficacy of larotrectinib in TRK fusion–positive cancers in adults and children. N Engl J Med 2018;378(8):731–9.

[11] Gopalakrishnan V, Spencer CN, Nezi L, Reuben A, Andrews MC, Karpinets TV, et al. Gut microbiome modulates response to anti-PD-1 immunotherapy in melanoma patients. Science 2018;359:97–103.

[12] Diouf B, Evans WE. Pharmacogenomics of vincristine-induced peripheral neuropathy: progress continues. Clin Pharmacol Ther 2019;105(2):315–7.

[13] Gage BF, Bass AR, Lin H, Woller SC, Steven SM, Al-Hammadi N, et al. Effect of genotype guided warfarin dosing on clinical events and anticoagulation control among patients undergoing hip or knee arthroplasty: the GIFT randomized clinical trial. JAMA 2017;318(12):1115–24.

[14] Rosenblat JD, Lee Y, McIntyre RS. Does pharmacogenomic testing improve clinical outcomes for major depressive disorder? A systematic review of clinical trials and cost-effectiveness studies. J Clin Psychiatry 2017;78(6):720–9.
[15] Tran AX, Ho TT, Gupta SV. Role of CVP2B6 pharmacogenomics in bupropion-mediated smoking cessation. J Clin Pharm Ther 2019;44:174–9.
[16] https://intermountainhealthcare.org/services/cancer-care/precision-genomics/pgx/. (Accessed March 24, 2019).
[17] Erzurumluoglu AM, Liu M, Jackson VE, Barnes D, Datta G, Melbourne CA, et al. Meta-analysis of up to 622,409 individuals identifies 40 novel smoking behavior associated genetic loci. Mol Psychiatry 2019;1–18. https://doi.org/10.1038/s41380-018-0313-0.
[18] McGinnis JM, Berwick DM, Daschle TA, Diaz A, Fineberg HV, Frist WH, et al. System strategy for better health throughout the life course. A vital direction for health and healthcare. In: Perspectives: expert voices in health and healthcare. Washington, DC: National Academy of Medicine; 2016. September 19.
[19] Richards S, Aziz N, Bale S, Bick D, Das S, Gastier-Foster J, et al. Standards and guidelines for the interpretation of sequence variants: a joint consensus recommendation of the American College of Medical Genetics and Genomics and the Association for Molecular Pathology. Genet Med 2015;17(5):405–24.
[20] Nair N, Camacho-Vanegas O, Rykunov D, Dashkoff M, Camacho SC, Schumacher CA, et al. Genomic analysis of uterine lavage fluid detects early endometrial cancers and reveals a prevalent landscape of driver mutations in women without histopathologic evidence of cancer: a prospective cross-sectional study. PLoS Med 2016;13(12):e1002206. https://doi.org/10.1371/journal.pmed.1002206.

Additional resources

- National Academies of Sciences, Engineering, and Medicine. Implementing and evaluating genomic screening programs in health care systems: proceedings of a workshop. Washington, DC: The National Academies Press; 2018. https://doi.org/10.17226/25048.
- National Academies of Sciences, Engineering, and Medicine. A framework for educating health professionals to address the social determinants of health. Washington, DC: The National Academies Press; 2016. https://doi.org/10.17226/21923.
- Magnan S. Social determinants of health 101 for health care: five plus five. NAM Perspectives. [Discussion paper]. Washington, DC: National Academy of Medicine; 2017. https://doi.org/10.31478/201710c. https://nam.edu/social-determinants-of-health-101-for-health-care-five-plus-five/.
- Clinical Pharmacogenetics Implementation Consortium (CPIC®). CPIC creates, curates, and posts evidence-based, peer-reviewed pharmacogenetic test clinical practice guidelines. https://www.Cpicpgx.org.
- PharmGKB. A compilation of genetic variants and gene-drug-disease relationships. https://www.pharmgkb.org.
- Davidson KW, Cheung YK, McGinn T, Wang YC. Expanding the role of N-of-1 trials in the precision medicine era: action priorities and practical considerations. NAM Perspectives. [Commentary]. Washington, DC: National Academy of Medicine; 2018. https://nam.edu/expanding-the-role-of-n-of-1-trials-in-the-precision-medicine-era-action-priorities-and-practical-considerations/.

- Murray MF, Evans JP, Angrist M, Chan K, Uhlmann W, Doyle DL, et al. A proposed approach for implementing genomics-based screening programs for healthy adults. NAM Perspectives. [Discussion paper]. Washington, DC: National Academy of Medicine; 2018. https://doi.org/10.31478/201812a. https://nam.edu/a-proposed-approach-for-implementing-genomics-based-screening-programs-for-healthy-adults/.
- National Academies of Sciences, Engineering, and Medicine. Enabling precision medicine: the role of genetics in clinical drug development: proceedings of a workshop. Washington, DC: The National Academies Press; 2017. https://doi.org/10.17226/24829.

IMPLICATIONS FOR EDUCATORS

Advances in technology have allowed us to make great strides in the understanding of health and disease. In the late 1800s, advances in microscopy led to dramatic growth in understanding of everything from bacteria to tissues and ideas such as the germ theory and Koch's postulates were developed. Advances in chemistry led to increased understanding of metabolites and respiration, and how perturbations result in disease states. This application of technological advances in other fields, applied to health and disease, has led to remarkable improvements in our ability to diagnose and treat a wide spectrum of diseases. Former scourges have disappeared in many parts of the world. Physicians have more diagnostic and treatment options available to them than at any time in history. This is incontrovertible.

However, the wisdom of the oft repeated admonition of Francis Peabody remains: "The fact must be accepted that one cannot expect to become a skillful practitioner of medicine in the four or five years allotted to the medical curriculum. Medicine is not a trade to be learned but a profession to be entered. All that the medical school can hope to do is supply the foundations on which to build" [1]. This perspective is as true today as it was a century plus ago and it will continue to be germane to the education of future physicians.

Medical educators are in an enviable position in that the end goal of the educational and training process is clear—to graduate physicians with the skills, knowledge and attributes required to provide the highest quality care throughout their professional careers. The challenge lies in the importance of understanding how clinical practice will look in the future. Will the roles and skills we identified as immutable remain as central to physicianhood? Will the advances in our scientific understanding and technology result in disruptive changes?

The visionary leaders who transformed medical education at the beginning of the 20th century did so based on trends in scientific discovery and technologic advancements despite not knowing or likely even imagining where they would lead. I would posit that we too are at a similar juncture. It is important that we consider the advances and external trends that will shape the practice of clinical medicine over the coming decades and then, based on our best judgment, engage in serious conversations regarding appropriate educational prerequisites and curricular revisions for optimal preparation of medical students for their future careers. The vision of what mid-21st century clinical medicine will be must inform how the medical education process and content evolves.

Given that it is unlikely that we will move to a system of admission to medical school directly out of high school, there are three distinct but interrelated steps to consider:

- Examination and reassessment of the foundational disciplines and subjects that are prerequisites for medical school applicants. This has far-reaching implications for the Academy and for all medical educators.
- Determination at the medical school level of what is essential, regardless of eventual specialization and career differentiation, what should be optional based on the career direction a student chooses, and what should be dropped from the curriculum.
- Faculty involved in resident education will need to envision what a physician practicing that discipline will do and then provide the requisite learning experiences to develop an independent level of expertise in the required knowledge and skills.

An overarching goal for all is to ensure that graduates and trainees are well-positioned to adapt as the clinical practice changes and that they have the scientific and pathophysiologic understanding, human to human "doctoring skills" and the intellectual curiosity and drive necessary to remain competent throughout their professional careers.

"We have to start by blowing up all these preconceptions about what the discipline is doing and that's going to take some real delicacy. I think we make a mistake if we move to too conceptual of a curriculum, which has been done many a times in medical education and nursing and pharmacy, where you back up and you teach these great concepts and you leave it to the learner to determine what are the pieces of information and things that they need. So blow up the preconceptions and restructure, teaching people not what they need to be today but we actually have to drive what they will be.

As long as we teach in a retrospective fashion about the jobs that already exist, then we're wasting a huge opportunity."

Patricia Hurn

Lest we underestimate the complexity and importance of this task, who could have anticipated 30 years ago that "minimally invasive" surgery would replace open surgery for so many procedures. That advances in pharmacotherapy, primarily statins, and catheter-based procedures would markedly decrease the number of open coronary artery bypass surgeries. That effective treatment protocols would have been developed for increasing numbers of malignancies and as a result, many patients now survive for decades. The list is long and the pace of change is accelerating.

As we embark on this journey there are a few caveats that must always be kept in mind.

Focus on the future professional. We must resist the desire to create the optimal curriculum and learning experiences for the current practice of clinical medicine.

Focus on learning. Faculty are oftentimes conflicted as their desires and needs no longer align with what is best for the medical students and residents. Service needs, faculty reluctance to learn new skills such as coaching rather than lecturing, and financial arrangements may conflict with optimal learning situations.

Telling is neither teaching nor learning. The temptation to revert to familiar behaviors on the part of the faculty is ever present. Ensuring that learners understand, integrate, and can apply knowledge and skills will require that faculty have an understanding of cognition and learning science that is incorporated into the educational programs.

Most importantly, Teach how to think, not what to think. Much of medical education and indeed the whole educational enterprise focus on what to think. Regurgitate the right answer, ace the test, impress the faculty with your command of the "facts." However, given the foreseen growth in scientific research and our understanding of health and disease, and the proliferation of tools and options for physicians, the focus must be on how to think. How to reason. How to frame problems and options.

We have already considered how the roles and skills that I have termed constant or immutable will need to be modified based upon readily identified trends. Let us now consider the three educational levels, premedical, medical and graduate medical education, in greater depth.

Reference

[1] Peabody FW. The care of the patient. JAMA 1927;88(12):877–82.

A new mental model for clinical education

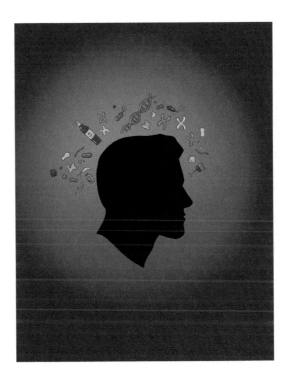

Public health measures, vaccinations, antibiotics, the advances in preventing and treating disease made over the last century have led to remarkable increases in longevity as well as in quality of life for millions.
However, the failure of what is often termed the "infectious disease
model" and the prevailing reductionist mindset to explain the complexity
of many modern maladies and chronic diseases is often rued. Alternative
models are not explicitly set forth for our students.

Implementing Biomedical Innovations into Health, Education, and Practice
https://doi.org/10.1016/B978-0-12-819620-5.00013-8

"shifting what has been largely a clinical care system that says what happens when you have a problem, to how it is that we can do everything imaginable in investing in the competencies and skills necessary to prevent those problems."

Mary Naylor

We are now poised to move from reactive to proactive medicine and we must ensure that our future physicians are trained accordingly. The confluence of technologic, analytic, and biologic advances in our understanding of the underpinnings of health and disease is making "health" a reachable goal. While the term "healthcare" has been broadly applied to clinical medicine for decades, it is a misnomer. The reality is that the principle focus has not been on "health." It has been on taking care of the sick. While care of the sick will continue to be an important and laudable calling, we need to provide our students the mental scaffolds upon which to build their knowledge that incorporates the factors that often precede manifest disease. Systems biology and its study of the interactions and networks across the life cycle, from developmental processes to physiologic responses through aging and senescence, provides a mental framework that accommodates current knowledge as well as the many gaps in our understanding. As set forth in the P4—predictive, personalized, preventive and participatory—model, every individual is a product of their inherent genomic information and environmental influences.

As expressed by Hood, "The combined influences of these factors…. generates each individual's phenotype—molecular, cellular and descriptive features—through networks of biological pathways that capture, transmit, and integrate signals and finally, send instructions to the molecular machines that execute the functions of life" [1].

We now are beginning to understand that we exist at the confluence of our genomic foundation, our microbiome, the physical and social environment in which we live and our behaviors (Fig. 13.1).

While the importance of genetics, behavior, and the physical and social environment has been recognized for decades and the contributions of the microbiome more recently, each of these areas is often considered in isolation rather than as part of the complete picture. We teach about the social determinants of health, about genomics, about the microbiome but the importance of the connections among these facets for future patients is not always apparent and is all too often ignored. While the ability to sequence the human genome has created a tremendous amount of interest and speculation regarding the role of our genome in determining our health, it is estimated that our genes contribute only 15–25% to our lifespan, suggesting that longevity has more to do with environment and behavior than with genetics [2]. Unfortunately, while genomics and the microbiome are readily

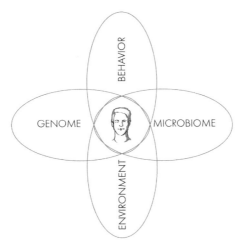

FIG. 13.1 Schematic depicting major factors that shape ones health and development of disease.

recognized as "medically relevant," environmental and behavioral factors are seen as somewhat extraneous and less germane to the students' and residents' future professional practices, resulting in lower learning priorities.

> "the intersection between the host and whatever environmental element that's causing disease will be better understood. The host characteristic will be outlined for everybody by that time, and I'm guessing full genome sequencing will just be part of the neonatal evaluation."
>
> *Clifford Lane*

While it is acknowledged that behavioral patterns account for approximately 40% of premature deaths and environmental factors for another 20% [3], the physiologic and pathophsyiologic linkages are just beginning to be understood. Epidemiologic studies have demonstrated that poverty, low educational attainment, and childhood circumstances such as malnutrition or physical and mental trauma, lead to adverse health outcomes that may only become evident years later. For example individuals who had experienced a stress related disorder (posttraumatic stress disorder, acute stress reaction, adjustment disorder, and others) were found to be at increased risk for developing autoimmune diseases [4]. It is well known that stress provokes an array of physiologic responses affecting the hypothalamic-pituitary-adrenocortical axis and the autonomic nervous system. However, the mechanisms remain to be elucidated.

There is growing evidence for the adverse effects of low grade environmental chemical exposures on physiologic systems. While primarily based on animal models, chemicals found in the environment have been shown to have a multitude of effects, including inducing changes that are passed to subsequent generations. The potential role of environmental chemicals in chronic disease and in societal problems such as the "obesity epidemic" has just begun to be explored. Likely additional deleterious effects of chemicals widely distributed in products, manufacturing processes and hence the environment will be discovered.

The built urban environment has long been recognized as a risk factor for pulmonary problems due to air pollution but now there is evidence that air pollution, particularly NO_2, is a risk factor for psychotic experiences among adolescents [5]. Quality green space has been found to be a stress reducer and contributes to mental well-being in children and adolescents as well as adults. It is hypothesized that low-grade inflammatory responses may be the physiologic link to disease.

Similarly, we are beginning to understand the multiple roles of the gut microbiome in health and disease. Because an individual's gut microbiome is determined largely by their environment, individuals who share households—both in the past and currently—have generally similar gut microbiomes. As expected, diet and lifestyle account for over 20% of gut microbiome diversity. Conversely, several studies suggest that less than 10% of the gut microbiota is heritable [6]. It is intriguing to consider that epidemiologic studies of diet and lifestyle may have their effect, at least in part, through modifications of the gut microbiome. While the full extent of the interactions between environmental factors and the microbiome also remain to be elucidated, the accumulating evidence supports the importance of this relationship.

Many major causes of morbidity and mortality are secondary to behavioral issues such as physical inactivity, substance abuse and addiction, tobacco abuse, and risk taking. Clearly an individual's behaviors and their environment are inexorably linked. Likewise, interactions between an individual's genome and their behaviors are being explored with intriguing results. A few examples include a meta-analysis of data that identified a region in chromosome 2 associated with aggressive behavior in children [7]. A study of almost 4000 individuals with a history of risky sexual behaviors and alcohol misuse identified a positive association with a single nucleotide polymorphism (SNP) in the *LHPP* gene [8]. And, a study found that variations in the *FBX045* gene were associated with emotional expression, a coping behavior for perceived stress [9]. In most of the recent studies of behavior and genomic associations, the caveat is the limitation in the ability to identify rare significant variants and that correlation does not necessarily equal causation. This area is just beginning to be researched and may result in the identification of genomic variants that put an individual at risk for behaviors which will ultimately have deleterious implications for their health.

Whether consciously or unconsciously, medical educators provide their students with the scaffolds upon which to build their knowledge of health and disease. As indicated in the Health-Disease Continuum Fig. 15.1, in Chapter 15, our students need a mental model of disease development that incorporates the concept that upstream intervention may prevent progression to disease in high risk individuals, may reverse presymptomatic disease with a return to health or at least ameliorate the development of clinical symptoms.

"I think we have to go from one-size-fits-all to a more individualized, or one might say personalized, or precision, approach to prevention that takes into account that particular individual's greatest risks, their behavior, their lifestyle.

I hope we're going to learn a lot in the next 20 years about this from the 'All of Us' program .... which is going to enroll a million Americans in a prospective longitudinal study collecting their electronic health records, all kinds of information about their health behaviors, diet and so on; both from questionnaires that they answer and wearable sensors that they walk around with that can also sample their environment. They will donate blood samples that will allow us to do all kinds of omics, certainly whole genome sequencing but also gene expression, proteomics, metabolomics, and so on. And with that cohort, which is going to be very geographically and ethnically diverse, and diverse in socio-economic status, gender and age, we'll finally have a really good idea about what correlates with good health outcomes and what doesn't.

This will allow a precision medicine approach to disease prevention and health maintenance. But it will also provide information about the most appropriate way to manage chronic illness so that a good result is achieved. Many of these participants will have been recruited because they are already enrolled in a health provider organization, so that as we learn what works, that knowledge can be rapidly implemented as a better practice for prevention as well as treatment of disease."

Francis Collins

The implications are noteworthy. The concept of a person or patient as a member of a group or population will be augmented or even replaced by the ability to approach each person as a unique individual—a dynamic, complex ecosystem that starts with their genomic foundation but then is shaped by their environment and behaviors; which in turn may be partially determined by their genomic makeup and their microbiome.

Disease states arise in the context of this unique ecosystem. The implications for therapeutic options will naturally encompass scientifically validated dietary and other non-pharmacologic "prescriptions." Disease risk will be expanded beyond the broad categories of age, gender, selected behaviors, existing conditions, and family history that we currently employ. At a minimum, the contributions of the person's genomic, microbiota, physical and social environment, and behaviors will be considered as they contribute to risk as well as to the actual development and progression of disease.

Consciously imparting such a mental model is essential for our students' and residents' abilities to integrate future scientific, technologic and analytic findings into their concepts of health and disease. While it is unlikely that they will be involved in interventions at each and every stage of the progression from normal to disease, equipping them to recognize when interventions have the potential to positively alter outcomes is of paramount importance if the imaginable goal of broadly envisioned "healthcare" is to be realized.

References

[1] Hood L, Friend SH. Predictive, personalized, preventive, participatory (P4) cancer medicine. Nat Rev Clin Oncol 2011;8(3):184–7.
[2] Kaplanis J, Gordon A, Shor T, Weissbrod O, Geiger D, Wahl M, et al. Quantitative analysis of population scale family trees with millions of relatives. Science 2018;360(6385):171–5. https://doi.org/10.1126/science.aam9309.
[3] McGinnis JM, Williams-Russo P, Knickman JR. The case for more active policy attention to health promotion. Health Aff (Millwood) 2002;21(2):78–93.
[4] Song H, Fang F, Tomasson G, Arnberg FK, Mataix-Cols D, Fernandezde la Cruz L, et al. Association of stress-related disorders with subsequent autoimmune disease. JAMA 2018;319(23):2388–400. https://doi.org/10.1001/jama.2018.7028.
[5] Newbury JB, Arseneault L, Beevers S, Kirwiroon N, Roberts S, Pariante CM, et al. Association of air pollution with psychotic experiences during adolescence. Published online March 27, 2019. https://doi.org/10.1001/jamapsychiatry.2019.0056.
[6] Rothschild D, Weissbrod O, Barkan E, Kurilshikov A, Korem T, Zeevi D, et al. Environment dominates over host genetics in shaping human gut microbiota. Nature 2018;555:210–5. https://doi.org/10.1038/nature25973.
[7] Pappa I, St Pourcain B, Benke K, Cavadin A, Hakulinen C, Nivard MG, et al. A genome wide approach to children's aggressive behavior: the EAGLE consortium. Am J Med Genet B 2015;171B:562–72.
[8] Polimanti R, Wang Q, Meda SA, Patel KT, Pearlson GD, Zhao H, et al. The interplay between risky sexual behaviors and alcohol dependence: genome-wide association and neuroimaging support for LHPP as a risk gene. Neuropsychopharmacology 2017;42:598–605. https://doi.org/10.1038/npp.2016.153.
[9] Shimanoe C, Hachiya T, Hara M, Nishida Y, Tanaka K, Sutoh Y, et al. A genome wide association study of coping behavior suggests FBXO45 is associated with emotional expression. Genes Brain Behav 2018;e12481. https://doi.org/10.1111/gbb/12481.

Additional resources

- National Academies of Sciences, Engineering, and Medicine. A framework for educating health professionals to address the social determinants of health. Washington, DC: The National Academies Press; 2016. https://doi.org/10.17226/21923.
- Brigham K, Johns MME. Predictive health. How we can reinvent medicine to extend our best years. New York, NY: Basic Books; 2012.

Premedical education

Premedical school matriculation requirements

The opening of Johns Hopkins Medical School in 1893 marked the first school built completely upon the pioneering model of a rigorous, experiential, laboratory-based study of medicine predicated on a foundation of basic medical sciences. Moreover, students were required to have a bachelor's degree as a prerequisite for admission—the most demanding admissions requirement of any medical school at the time [1].

The success of Johns Hopkins and the other schools that subsequently embraced the science-based curriculum was such that by the time Abraham Flexner was commissioned in the early 1900s by the Carnegie Foundation to survey the state of medical education in the United States and Canada, he wrote,

> ... the privilege of the medical school can no longer be open to casual strollers from the highway. It is necessary to install a doorkeeper who will, by critical scrutiny, ascertain the fitness of the applicant. [2]

Implementing Biomedical Innovations into Health, Education, and Practice
https://doi.org/10.1016/B978-0-12-819620-5.00014-X

When contrasting medical school with college, Flexner went on to say,

> Now the medical sciences proper – anatomy, physiology, pathology, pharmacology – already crowd the two years of the curriculum that can be assigned to them; and in so doing, take for granted the more fundamental sciences – biology, physics, and chemistry, for which there is thus no adequate opportunity within the medical school proper. ... these conclusions emerge: by the very nature of the case, admission to a really modern medical school must at the very least depend on a competent knowledge of chemistry, biology, and physics. [3]

And so, enshrined in the 1910 Flexner report was the strong argument for antecedent preparation in biology, chemistry and physics as prerequisites for admission to medical school. The logic upon which these recommendations rested was twofold—primarily that they were necessary to understand the medical sciences as presented in the first two basic science years, but secondarily to "weed out" intellectually unqualified students during their premedical training.

Focusing first on the role of premedical requirements as an intellectual foundation, it was envisioned by Flexner that matriculating medical students would be immersed in the relevant cutting-edge basic science courses and accompanying laboratories. Through their coursework, and their experiences in the laboratories, the students would be given the opportunity to see for themselves the validity of the "new ideas" that scientific discovery was enabling. The Einstein quote "learning is experience, everything else is just information" is applicable. Through experience, the students of that era gained knowledge about the most advanced science and how it was relevant to their understanding and potentially, to their future clinical practices. But to fully comprehend what they were being taught in medical school required a foundational knowledge base. Hence, the recommended pre-matriculation requirements.

Surprisingly, despite dramatic changes in the scientific bases for medicine and recurring calls for reform, the question of how best to prepare students through their premedical curriculum for the rigors of a medical education as well as for a lifetime of professional practice has not resulted in widespread changes [4]. Although the content has evolved as scientific knowledge has progressed, the required courses identified over a century ago remain. Moreover, rather than encouraging the use of the most sophisticated and creative techniques to validate recent scientific discoveries, often the premedical laboratory requirements instruct students to replicate tried-and-true experiments, or even "dry lab" them, which results in mundane and dulling, rather than intellectually stimulating, experiences.

As identified by Flexner, the second function of pre-matriculation requirements was to "weed out" intellectually unqualified students. This role continues. As expressed by Brenner when advocating for chemistry courses tailored to the needs of biomedical students, "Chemistry is the foundational discipline for molecular medicine. Moreover, chemistry departments are of key importance in educating premedical students and in filtering out those students who are insufficiently disciplined or intellectually inclined to grasp the scientific basis for medicine" [5]. Unfortunately, not only are "unqualified" premedical students discouraged from pursuing application to medical school, there are additional untoward consequences.

While it is well recognized that medicine needs to enhance the numbers of underrepresented minority students in medicine, the negative impact of using chemistry courses to "assess" students who might potentially be excellent physicians, has not been adequately considered. Studies of underrepresented minority and women undergraduates at highly competitive public and private universities show that their experiences in chemistry courses are the number one reason for abandoning their interest in pursuing a career in medicine [6, 7].

Although premedical requirements vary among medical schools, they generally include 1 year of biology, 1 year of physics, 1 year of English, and 2 years of chemistry—including organic and sometimes physical chemistry, and oftentimes calculus. However, it has long been recognized that in many colleges these prerequisites are only peripherally applicable to human biology and are not fulfilling their purpose of preparing students upon matriculation into medical school for the sciences they will encounter. In addition, the heavy emphasis of science-based premed requirements is seen as undermining the principles of a broad-based education.

In response, the 2009 AAMC—HHMI Scientific Foundations for Future Physicians report eschewed required courses and instead enumerated competencies, leaving it to the colleges as to how best they would be met. Eight competencies focusing on the scientific foundations enabling future physicians to practice science-based medicine, with learning objectives and examples accompanying each competency, were developed.

1. Apply quantitative reasoning and appropriate mathematics to describe or explain phenomena in the natural world.
2. Demonstrate understanding of the process of scientific inquiry and explain how scientific knowledge is discovered and validated.
3. Demonstrate knowledge of basic physical principles and their applications to the understanding of living systems.
4. Demonstrate knowledge of basic principles of chemistry and some of their applications to the understanding of living systems.
5. Demonstrate knowledge of how biomolecules contribute to the structure and function of cells.
6. Apply understanding of principles of how molecular and cell assemblies, organs, and organisms develop structure and carryout function.

IV. Implications for educators

7. Explain how organisms sense and control their internal environment and how they respond to external change.
8. Demonstrate an understanding of how the organizing principles of evolution by natural selection explain the diversity of life on earth [8].

A parallel report, Behavioral and Social Science Foundations for Future Physicians [9], was released in 2011. It adopted and merged the findings from the 2004 report from the Institute of Medicine "Enhancing the Social and Behavioral Science Content of Medical Schools Curricula" and the 2005 Royal College of Physicians and Surgeons of Canada report "The Can Meds 2005 Physician Competency Framework." The behavioral and social science domains taken from the Institute of Medicine report were: patient behavior, mind-body interactions in health and disease, physician role and behavior, physician-patient interactions, health policy and economics, and social and cultural issues in healthcare. These domains were matrixed with the physician roles from the Royal College of Physicians and Surgeons report: professionals, communicators, collaborators, managers, health advocates, scholars, and medical experts. Specific objectives and demonstrable actions were then developed. Although it did not explicitly identify premed competencies, the report referenced the importance of attention to the social and behavioral sciences as part of premedical school preparation.

Recognizing that the practice of medicine and its scientific underpinnings are in a time of great change, extensive deliberations have resulted in further refinements to the desired premedical competencies as listed in Table 14.1. Compared to a list of required courses, this provides premed students greater clarity for undergraduate experiences and educational preparation. However, even if these competencies establish an optimal foundation for future clinical realities, in the absence of agreed upon

TABLE 14.1 Core competencies for entering medical students

- *Service orientation:* Demonstrates a desire to help others and sensitivity to others' needs and feelings; demonstrates a desire to alleviate others' distress; recognizes and acts on his/her responsibilities to society; locally, nationally, and globally.

- *Social skills:* Demonstrates an awareness of others' needs, goals, feelings, and the ways that social and behavioral cues affect peoples' interactions and behaviors; adjusts behaviors appropriately in response to these cues; treats others with respect.

- *Cultural competence:* Demonstrates knowledge of socio-cultural factors that affect interactions and behaviors; shows an appreciation and respect for multiple dimensions of diversity; recognizes and acts on the obligation to inform one's own judgment; engages diverse and competing perspectives as a resource for learning, citizenship, and work; recognizes and appropriately addresses bias in themselves and others; interacts effectively with people from diverse backgrounds.

TABLE 14.1   Core competencies for entering medical students—cont'd

- *Teamwork:* Works collaboratively with others to achieve shared goals; shares information and knowledge with others and provides feedback; puts team goals ahead of individual goals.

- *Oral communication:* Effectively conveys information to others using spoken words and sentences; listens effectively; recognizes potential communication barriers and adjusts approach or clarifies information as needed.

- *Ethical responsibility to self and others:* Behaves in an honest and ethical manner; cultivates personal and academic integrity; adheres to ethical principles and follows rules and procedures; resists peer pressure to engage in unethical behavior and encourages others to behave in honest and ethical ways; develops and demonstrates ethical and moral reasoning.

- *Reliability and dependability:* Consistently fulfills obligations in a timely and satisfactory manner; takes responsibility for personal actions and performance.

- *Resilience and adaptability:* Demonstrates tolerance of stressful or changing environments or situations and adapts effectively to them; is persistent, even under difficult situations; recovers from setbacks.

- *Capacity for improvement:* Sets goals for continuous improvement and for learning new concepts and skills; engages in reflective practice for improvement; solicits and responds appropriately to feedback.

Thinking and reasoning competencies

- *Critical thinking:* Uses logic and reasoning to identify the strengths and weaknesses of alternative solutions, conclusions, or approaches to problems.

- *Quantitative reasoning:* Applies quantitative reasoning and appropriate mathematics to describe or explain phenomena in the natural world.

- *Scientific inquiry:* Applies knowledge of the scientific process to integrate and synthesize information, solve problems and formulate research questions and hypotheses; is facile in the language of the sciences and uses it to participate in the discourse of science and explain how scientific knowledge is discovered and validated.

- *Written communication:* Effectively conveys information to others using written words and sentences.

Science competencies

- *Living systems:* Applies knowledge and skill in the natural sciences to solve problems related to molecular and macro systems including biomolecules, molecules, cells, and organs.

- *Human behavior:* Applies knowledge of the self, others, and social systems to solve problems related to the psychological, socio-cultural, and biological factors that influence health and well-being.

Endorsed by the AAMC Group on Student Affairs Committee on Admissions. Reproduced from students-residents.aamc.org/applying-medical-school/article/core-competencies/ [Accessed 29 March 2018]. ©2019 Association of American Medical Colleges, with permission. All rights reserved.

IV. Implications for educators

assessment tools, metrics and standards, it is not particularly meaningful when an undergraduate program asserts that a student has demonstrated "satisfactory achievement" of the competencies. Developing competencies rather than course lists is a step in the right direction but absent assessment criteria and valid tools, the default too frequently is specific courses.

Developing appropriate premedical requirements is complex. Just as the general outlines of today's requirements were introduced to prepare future medical students to take full advantage of the most rigorous and "modern" approach to clinical medicine in the late 19th and early 20th centuries, so too, they must be updated and reworked to fulfill this purpose in the 21st century. Current transformative technologies are those that enable rapid gene sequencing, the measurement of a myriad of proteins, noninvasive sensors allowing measurement of physiologic variables and behavioral parameters, advances in understanding cell signaling, genomic editing and computing power to analyze massive amounts of aggregated data. What are the educational foundations needed by medical students to fully engage the medical curriculum and have the foundation for a lifelong professional career? What are the learning experiences that will enable using the data generated from technologic advances? But as discussed in Section II, Constants in Medicine, it is the human to human interaction that is the essence of medicine. Thus, the educational foundation for medicine is not solely based on science and technology and must be balanced with the humanities.

The end goal of the educational and training process—to graduate physicians with the skills, knowledge and attributes required to provide the highest quality care throughout their professional careers—should drive the process of determining premedical requirements.

"Focus on learning how to learn. That means you need to become very facile with how you utilize computational approaches for whatever question you're interested in. Constantly assess how you evaluate the validity of information, and how you decide what's real evidence and what is fake news. It's going to get harder and harder to tell the difference, given the proliferation of sources of information. You have to think quantitatively and you have to be prepared, if you're going to be a physician, to know where to find the trustworthy evidence that you're going to need to take good care of patients.

But on top of that, and I'll come back to this because I've been talking about all this high-tech stuff. You've got to decide whether you really love people. Do you really want to give a big part of your life, not to making your own self better, but to helping other people? Is that a calling that you feel strongly you want to take part in? Are you prepared?—because it's going to require you to make sacrifices in terms of your own ease and enjoyment of life?"

Francis Collins

As educators consider the prerequisite knowledge base for matriculation into medical school, an important consideration is the fact that many students who initially express an interest in pursuing medicine as a career end up working in other disciplines and professions. Therefore, premed requirements should also focus on knowledge and skills that transcend the context of medicine and are integral to a well-rounded education.

In response to the question "If you had a high school student come to you and say, 'What should I do to prepare myself for the world you are making for me?'"

"I would say do what is most interesting for you academically although recognizing that there are entry requirements currently for medical schools, some of which don't make sense. We're trying to change those, I mean having to take calculus but not statistics, how ridiculous is that; and think in your training about systems. Whether you're in biology or anthropology, try and develop a systems view of the world. So you don't want to become a doctor that when someone talks to about population health you think of your 2000 patients on your panel, you want to think of yourself as one of 850,000 physicians in a pool with 3 million nurses, 5500 hospitals, with $3.2 trillion in your pocket to deploy for the optimal outcome of 320 million Americans and to stay focused on that bigger picture."

James Madara

Based on my assessment of how the practice of medicine will evolve, on the following pages I suggest for consideration examples of premed requirements. Some of the examples are closely related to existing requirements and modest modifications would align them with desired competency goals. In other situations, new courses and experiences are needed. It is likely that advances in computer-based instructional technology will allow some prerequisites to be taught through online courses that would ensure a baseline level of shared understanding by all medical school matriculants. This is clearly not consistent with the current business model of schools and colleges. While disruptive, the advantages would be great.

Medical and scientific terminology

A recurring admonition for physicians is to refrain from medical jargon when speaking with patients and their families and to make certain to use terms readily understood by the lay public. Entering medical students are no different. The language of biomedical science and clinical medicine often includes terminology, jargon, and acronyms that are not part of the students' everyday vocabulary. Ensuring a common level of understanding of frequently encountered terms, prefixes, suffixes, and the like would ease the transition into medical school.

Rather than expecting colleges and universities to mount new courses on medical terminology and jargon, online courses requiring students to demonstrate mastery of the material prior to application or matriculation into medical school should be considered. These already exist [10]. While some are designed to enhance the lay public's understanding of medical terms, others are focused on developing a professional's vocabulary. Students could readily work in a self-paced independent study mode and demonstrate proficiency through an online assessment which would prepare them for the words they will be encountering in medical school.

Numeracy

Mathematical literacy, or numeracy, is essential. Whether reading the medical and scientific literature, understanding how normal and abnormal laboratory test categorizations are derived, or interpreting the results of data analytics, an understanding of statistics is absolutely necessary. Developing an awareness of the appropriate and inappropriate uses of common statistical measures, knowing the meaning of common terms (e.g., risk reduction, relative risk, absolute risk, false positive, false negative), and understanding Bayesian statistics and how to calculate positive and negative predictive values is essential for premedical students. Given the current realities of clinical medicine and biomedical science, and the expectation that the availability of data will continue to increase markedly, it is important that students possess the knowledge to interpret numerical data. Biomedical statistics are as much a part of communicating meaning as are the words used in medicine and in the life sciences.

I would argue that numeracy, specifically a basic understanding of statistics, is essential to the education of all undergraduate students. To be an informed citizen an understanding of how statistics are manipulated to support a narrative by the media, marketing firms, and others is necessary.

A universal premed "math" requirement should be a demonstrable understanding of biomedical statistics. This will provide future physicians with a much better educational foundation than, for example, the traditional medical school prerequisite calculus course(s).

Genetics, epigenetics, and molecular biology

In addition to preparing the premedical student for medical school from a "medical and scientific terminology" and "mathematical literacy" perspective, a rigorous understanding of genetics (the language of life), epigenetics, and molecular biology is essential. While this is the focus of many biology courses, it should be a requirement for premedical students. Whether it is genetically modified foods, gene editing to cure otherwise fatal diseases, or using genetically altered organisms to produce therapeutic agents, organs or structures for transplant into humans, genetics is at the nexus of many current debates.

Understanding heritability patterns, how genes are activated at different stages of an organism's development and life cycle, how environmental factors may lead to epigenetic changes, the realities of genetic editing, and how genomic variants may result in disease is useful information for anyone wishing to be conversant with many current issues. As this is an area of remarkable progress, the necessary level of understanding will require frequent updates and revisions. In my opinion, this too should be part of the education for all college students. A basic grasp of genetics is necessary to be an informed citizen.

"… fundamental for education is understanding genetics. I mean basic concepts of what is heterozygous, what is autosomal dominant, what is recessive, X linked, ….This is logic, this is not memorization. This is just logically thinking about the fact that you have two chromosomes, that they're actually segregating, that there is in fact recombination, that there are mosaics and those mosaics are relevant to both common diseases as well as relevant to cancer. That's basic fundamentals; being able to understand segregation, understand recombination, and understand this in the context of phenotypes."

Evan Eichler

Molecular biology, providing a foundation of understanding of the molecular interactions within cells, is also appropriate preparation for premedical students. However, unlike numeracy, genetics and epigenetics it may be less germane for undergraduates who are not contemplating a career in science or medicine.

Computer programming as a "foreign language"

While foreign languages are generally not required for application to medical school, they are frequently listed as an undergraduate requirement. The argument is made that learning a foreign language enhances one's understanding of different cultures and broadens one's worldview. That said, an introductory course to computer programming would be valuable for premedical students, and I would suggest, all students, to familiarize them with the universal language of computers. The goal would not be to master coding but rather to understand the logic and approach used by programmers. Becoming familiar with concepts such as abstraction or the ability to develop a simplified model of a more complex process and "divide and conquer"—to break a complex problem into pieces and then consolidate the solutions—will provide knowledge and insight into the world of computer technology that permeates our lives.

Logic

The discipline of thought, developed through a logic course, is applicable to informed understanding of a wide range of information from scientific to political arguments. Identifying and learning to avoid common errors in thinking; assessing the validity and cogency of arguments in realms such as science, discussion of ethical and moral issues, and law; and developing the ability to construct a valid argument, reason thoughtfully and recognize a non sequitur are broadly relevant skills that should be a prerequisite for future physicians.

As medical students, they will be involved in discussions with faculty, peers, patients and others on legal, ethical, and scientific topics. The scope and implications of these discussions will increase as they progress in their professional careers. Members of the public will look to them for advice and perspective. Physician-assisted suicide, chimeric organisms, gun violence, the clinical definition of death, these are but some of the current topics debated within the medical community and society at large. The ethical and legal implications of technologic monitoring, the balance of the physician's responsibilities to their individual patient versus societal obligations and the responsible use of genetic editing are among the issues they will likely face during their professional careers. Rigor in thinking and the ability to develop cogent arguments are foundational skills that should be developed as part of their premedical education requirements.

Communication

Communication is a fundamental skill, not only for premedical students, but for all students. In view of the growing reality of a "post text" world, the traditional collegiate emphasis on written communication should be expanded to include interpersonal, formal and informal verbal communication with an overt emphasis on the importance of knowing one's audience. Incorporating rhetoric, improvisation, and the use of technologies such as video should be considered. Involving students in theater where vocal technique and similar approaches are taught could be a powerful communication learning experience.

An often overlooked aspect of communication is the importance of listening. Practicing active listening and conscious attention to nonverbal communication, while introducing students to the concepts of body language and clues that nonverbal communication provides regarding the veracity of the statements an individual is making, are widely applicable skills. If such a course is developed specifically for premedical students, it should include advocacy, research and professional communication.

Understanding the human condition

"… it's going to be a different future and I think in the undergraduate years that learning psychology or some other way of relationship development and building empathy (should be required).

I mean think about the world, that's one thing that given the human condition we shouldn't be expecting technology to provide, we need empathy.

So finding ways in the undergraduate years to expose people to this; it could be things like community service, structuring it in a way that was part of the requirements."

Gary Kaplan

While it is likely that the diagnostic and therapeutic options available to physicians will change and the context and means of interacting with patients will expand, the interaction with individuals who are vulnerable, facing life crises, dealing with disabilities and disease, and often contemplating difficult decisions, will remain a constant. Most premedical students have not had personal experience with the pathos of human existence even though disease and tragedy may have been experienced by elderly relatives or occasionally by friends and acquaintances. A challenge then is not just to expose students, as through newscasts or staged educational videos, but rather to emotionally engage them. As future physicians, they will need to develop empathy for the broad spectrum of emotions—the sense of betrayal, the loss of identity, the helplessness, the despair—felt by patients and their families when faced with devastating conditions and recognize that even in the most dire circumstances the human spirit can show incredible resilience, strength, and courage.

As expressed by Gunderman and Kanter "We can perceive only what our mental models enable us to see and imagine. If these models have not been broadened and deepened by contact with the great poets, philosophers, and prophets of human civilization, then our outlook will be narrower and more myopic than it needs to be…. If the physicians of tomorrow are to serve the profession of medicine, their communities, and their patients, it is vital that today's medical students arrive having reflected broadly and deeply on what it means to be human" [11].

Literature, art, theater, video, and music are often utilized to provide a window on the human condition. Incorporating co-creation of literary, musical, and/or artistic works based on the stories told to the students by individuals struggling with chronic disease, disabilities, or addiction, would invest the students in the experience and likely enhance their appreciation for the emotional and psychological impact experienced by the patients and families. Such a learning experience could be very powerful and would provide students with insight into the impact of disease on individuals affected directly and indirectly. Development of these courses will require faculty to be willing to experiment with creative approaches beyond the traditional teaching of the humanities, arts and psychology.

What to eliminate?

Elimination of existing programs or making changes to established requirements is among the most difficult responsibilities of the Academy. There are always "thoughtful" arguments for maintaining the status quo, with perhaps only minor modifications or, preferably, additions to existing requirements. Nothing is broken, why change it! This resistance to change is understandable. Privately, college administrators and leaders worry about how to sustain their faculties in departments such as physics and chemistry if those courses are no longer prerequisites for medical school as there are insufficient numbers of students pursuing majors in these departments to justify the existing faculty size.

Going forward, the focus of premedical requirements must be on the optimal preparation of students for medical school and their professional practice. The shift toward competency-based educational requirements is progress. However, the challenge to develop consensus regarding competencies, appropriate and accepted assessments and performance standards to demonstrate achievement of competence remains.

I would advocate for using an approach equivalent to "zero-based budgeting." Whether it be competencies, or as an interim step courses/disciplines, the starting assumption should be that it would not be a pre-matriculation requirement. Using this logic, unless a compelling rationale emerges for their importance to the professional function of the future physician, the following prerequisite courses should be eliminated: physics, general biology, chemistry including organic chemistry and physical chemistry, English, and calculus.

Clearly, this is a departure from what has been in place for decades and will thus be controversial. Rather than arguing about the importance of existing premedical course requirements, the starting point for discussion should be the end goal of the continuum of medical education and the foundation students need for their future professional careers. Then, the experiences and courses that best develop that foundation can be articulated. As the practice of clinical medicine continues to evolve, a periodic re-examination of pre-matriculation competencies, assessments, and metrics will be essential. It is our responsibility as medical educators to ensure that "new" premedical requirements do not deviate from the needs for which they were developed.

Medical college admissions test (MCAT)

Given the heterogeneity of prerequisite courses, the reality of grade inflation and the difficulty in determining the actual level of achievement for a given grade across colleges and universities, the MCAT has emerged as the one tool that provides a uniform assessment of students' knowledge. A comprehensive review and revision of the MCAT resulted in a new test that was first introduced in 2015 [12]. The blueprint for the revised MCAT sought to balance a broad range of "competencies" in the four sections, Biological and Biochemical Foundations of Living Systems; Chemical and Physical Foundations of Biological Systems; Psychological, Social, and Biological Foundations of Behavior; and Critical Analysis and Reasoning Skills [13] to provide as comprehensive as possible an assessment of the students' intellectual preparation for medical school.

However, the widespread use of professional preparatory courses designed solely to enhance MCAT performance, and some that even "guarantee" achieving an advertised score or better, undermines its utility as a broad-based assessment of students' knowledge. Moreover, as students realize that their admission to a medical school may hinge on their MCAT score, the anxiety it engenders is immense. While I fully realize that a level of intellectual preparation and achievement is necessary for future physicians, the reliance on the MCAT as a determinant of suitability for admission to medical school should be reconsidered and potentially eliminated, or used only to establish a minimum threshold that has been studied and shown to correlate with successful completion of medical school.

The ultimate goal with the establishment of agreed upon premedical competency requirements is that demonstration of competency in all prerequisites would eliminate the need for a national single point in time test such as the MCAT.

Questions for reflection

In the 21st century what should be the role of premedical matriculation requirements?
Is it time to rethink the purpose of premedical education requirements as part of the continuum of medical education and if so what should be the components thereof?
Should the "information transfer" or didactic portions of premedical competency requirements be web-based with undergraduate programs focused on the experiential aspects of learning?

References

[1] Ludmerer KM. Learning to heal: the development of American medical education. New York: Basic Books; 1985. p. 58–60.

[2] Flexner A. Medical education in the United States and Canada: a report to the Carnegie Foundation for the advancement of teaching. [Bulletin number four]. Boston, MA: Updyke; 1910. p. 22.

[3] Flexner A. Medical education in the United States and Canada: a report to the Carnegie Foundation for the advancement of teaching. [Bulletin number four]. Boston, MA: Updyke; 1910. p. 24–5.

[4] Chambers DA, Cohen RL, Girotti J. A century of premedical education. Perspect Biol Med 2011;54(1):17–23.

[5] Brenner C. Rethinking premedical and health professional curricula in light of MCAT 2015. J Chem Educ 2013;90:807–12.

[6] Barr DA, Matsui J, Wanat SF, Gonzales ME. Chemistry courses as the turning point for premedical students. Adv Health Sci Educ 2010;15:45–54.

[7] Barr DA, Gonzales M, Wanat SF. The leaky pipeline: factors associated with early decline in interest in premedical studies among underrepresented minority undergraduate students. Acad Med 2008;83:503–11.

[8] Association of American Medical Colleges—Howard Hughes Medical Institute. Scientific foundations for future physicians. Washington, DC: AAMC; 2009. http://www.aamc.org/scientificfoundations.

[9] Association of American Medical Colleges. Behavioral and social science foundations for future physicians; 2011. https://www.aamc.org/download/271020/data/behavioralandsocialsciencefoundationsforfuturephysicians.pdf.

[10] Understanding medical words: a tutorial from the National Library of Medicine. https://medlineplus.gov/medicalwords.html.

[11] Gunderman RL, Kanter SL. How to fix the premedical curriculum: revisited. Acad Med 2008;83(12):1158–61 [p. 1161].

[12] Schwartzstein RM, Rosenfeld GC, Hilborn R, Oyewole SH, Mitchell K. Redesigning the MCAT exam: balancing multiple perspectives. Acad Med 2013;88:560–7.

[13] What's on the MCAT exam? https://aamc-orange.global.ssl.fastly.net/production/media/filer_public/44/e8/44e8b9aa-5000-490c-8a6a-c7ff8d01874d/combined_mcat-content_new_013118.pdf. (Accessed March 29, 2019).

Additional resources

- National Academies of Sciences, Engineering, and Medicine. Data science for undergraduates: opportunities and options. Washington, DC: The National Academies Press; 2018. https://doi.org/10.17226/25104.
- Institute of Medicine. Improving medical education: enhancing the behavioral and social science content of medical school curricula. Washington, DC: The National Academies Press; 2004. https://doi.org/10.17226/10956.
- Frank JR, editor. The CanMEDS 2005 physician competency framework: better standards, better physicians, better care. Royal College of Physicians and Surgeons of Canada; 2005.

Medical school education

"I think that medical student education is going to need to be dramatically different than memorizing theories and facts, many of which will later be found out to be wrong. The entire medical education process will need to be built upon internet-based up-to-date information that is constantly changing. And the goal of medical education will not be to have students have a vast fund of knowledge of things in their heads that they can produce on a test, but rather the ability to think through complicated situations and to know where to find information when you need it."

Francis Collins

Implementing Biomedical Innovations into Health, Education, and Practice
https://doi.org/10.1016/B978-0-12-819620-5.00015-1

Although the general structure of a four-year medical student educational experience has been in place for over a century, the content and context of medical student education is an ever-evolving field. As our understanding of the scientific bases of disease changes, it becomes incorporated into students' education. As the diagnostic and therapeutic options available to physicians change, medical students are immersed in clinical settings where these modalities are applied. As the site of care delivery changes, the educational venues in which medical students learn also change. And, as discussed in Chapter 19 as advances in the understanding of cognition and how we learn occur, they are translated into the process of medical education. Going forward, it is essential that medical educators think comprehensively about the totality of medical student education—content, context and process.

As the Flexner report, funded in 1910 by the Carnegie Foundation had profound effects on today's medical education, so too the Carnegie study commissioned at the beginning of the 21st-century must be considered. Through in-depth study of 14 institutions including academic medical centers and nonuniversity affiliated teaching hospitals, hundreds of interviews and focus groups, recommendations for reform of medical education in this century were developed. The results are set forth in the 2010 report *Educating Physicians: A Call for Reform of Medical School and Residency* [1].

The study recommends a medical education system that:

- Maximizes flexibility in the process of achieving standardize outcomes.

   Rather than focusing on time and process requirements, standardize learning outcomes and competencies but allow for individualized paths for fulfillment.
- Creates opportunities for integrative and collaborative learning.

   Physicians integrate knowledge and skills from multiple domains and disciplines in their daily functions. Yet "formal" courses and clinical experiences frequently are single discipline or domain based. The learning of basic, clinical and social sciences should be integrated to facilitate understanding of how they relate to patient care and the physicians' roles within the clinical environment.
- Inculcates habits of inquiry improvement.

   To promote a commitment to excellence throughout one's professional career requires developing habits of inquiry and improvement. The educational process emphasizing factual knowledge does not foster the curiosity and drive for improvement necessary for continued professional excellence. Thoughtful medical student and resident engagement in inquiry, discovery and innovation should be supported by all academic medical centers.

- Provides a supportive learning environment for the professional formation of students and residents.

The development of the future physicians' professional identities, including professional values, actions and aspirations is too frequently neglected in teaching and assessment practices. The development of compassion, empathy and humane practice should be an aspiration of all educational programs.

Echoing the recommendations of the 1910 study, the 2010 report recommended that changes in the practice of clinical medicine and advances in learning sciences again require major reforms in our medical education system. Noteworthy is the emphasis on process and reaffirmation of the importance of professionalism and standards of excellence. The challenges outlined by the report are recognized by most medical educators and the recommendations are considered to be sound and worthy of implementation. As always, the real challenge lies in the implementation and this should not be underestimated.

Scope of practice

Complementary to the proposals set forth in *Educating Physicians: A Call for Reform of Medical School and Residency* are the recommendations that follow in this chapter. While it is impossible to predict the pace of change in the practice of clinical medicine, fundamental for the profession and the Academy is the answer to the question: What should be the role of physicians? Caring for the sick, alleviating pain and suffering, rehabilitating the infirm and disabled and providing compassionate end of life care is a noble undertaking. This has been the province of physicians for centuries. Indeed, while we use the terms "healthcare" and "health systems" they are misnomers when considering what is actually the focus of hospitals, a majority of clinics, and the physicians, nurses and other professionals working in "healthcare." The reality is that medicine is primarily about disease care. While the ideal of preventing disease and maintaining health is a goal, the reality is that the vast majority of physician time and focus is dedicated to curing or ameliorating disease.

As seen in Fig. 15.1, there is a continuum from health to disease. Physicians traditionally have focused on the late early and late disease stages. As we understand more about the underpinnings of disease causation, the promise of being able to maintain health by targeting interventions in healthy individuals at risk to develop specific diseases, the normal high risk stage, will increasingly be realized. A question facing the profession of medicine is whether physicians should enlarge their scope of practice to include specific interventions in healthy individuals to minimize their likelihood of developing diseases for which they are at risk, or if physicians should

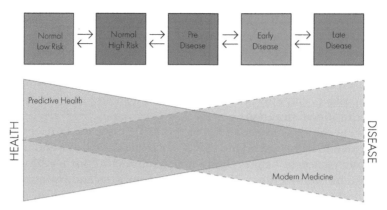

FIG. 15.1   The health-disease continuum.

remain primarily disease focused but expand their practice to include intervening earlier in the course of disease prior to symptoms developing—the pre-disease stage—in the hope of restoring the individual to full health. An open, thoughtful debate rather than a laissez faire approach to the role and focus of physicians is needed.

Clearly, this discussion directly impacts medical student and resident education. It will further define the knowledge and skills needed by future physicians and it may result in the creation of new specialties.

Irrespective of the outcome regarding physician scope of practice at the national level, at the academic health system level medical educators are facing decisions that will have profound implications not only for the educational process but also their students' and trainees' professional careers and society at large. Their decisions will influence the scope of professional practice.

Model of disease

Based on advances in biomedical sciences, a reconceptualization is needed of the model of health and disease that is used to situate medical students' and residents' learning. As implicitly and explicitly this model defines what is deemed relevant the importance of this model as the guide for medical education cannot be overemphasized.

While the appropriateness of a biopsychosocial approach to patient care has been acknowledged for decades, medical education continues to be dominated by a pathophysiologic focus built upon the infectious disease model and the concept of Occam's razor. While appealing in its simplicity, the reality is that many of the maladies currently accounting for the majority of human morbidity and disease are multifactorial and complex.

Students must be introduced, right from the beginning of their professional education, to a model of health and disease that will serve as a mental scaffold upon which to build their understandings not only of pathophysiology but also the information required to understand an individual's risk for disease, health and disease status and the multiple professional disciplines that will "care" for the individual throughout their life journey.

Such a model, as proposed in Chapter 13 positions each of us as existing at the intersection of our genetic makeup, our behaviors, our environment which includes the social determinants of health and our microbiota. (Notably, it is thought that medical care contributes only about 10% to an individual's health.) Just as the germ theory supplanted concepts of marasmus and imbalance in humors, so too the advances in our understanding of the complex interrelationships among these factors suggest that a balanced ecosystem model of health is more appropriate and that imbalance in the ecosystem results in symptomatic disease. Interestingly, while the components are identified differently, the concept that health is dependent upon inherent predilections that manifest as illness only when other events occur is an old concept that has taken different forms in different cultures throughout history.

A simplistic example would be educating students that even for infectious diseases, while the most proximate cause is a specific pathogen, the totality of the individual's ecosystem sets the stage for the pathogen to be manifest. For example, *Clostridioides difficile* infection, initially called antibiotic associated colitis, arises in individuals with antecedent antibiotic, chemotherapy or immunocompromised states. Oftentimes it is seen in hospitalized patients where colonization with *C. difficile* too frequently occurs. The focus for years was on eradicating the *C. difficile* through antibiotic therapy. Unfortunately, while therapy usually resulted in clinical resolution of symptoms, recurrence, sometimes within days of cessation of antibiotic therapy, was experienced by about 25% of patients, with additional recurrences in about 40% of these patients [2]. This approach, targeting the offending bacteria, is a perfectly logical response when one conceptualizes the problem as an isolated offending organism or pathogen. Eradication of that pathogen should lead to "curing the disease." However, as our understanding of the complex relationships within the gut microbiota has developed, we now know that the therapies antecedent to the development of *C. difficile* colitis, radically alter the gut microbiome allowing space for *C. difficile* to proliferate and manifest as disease. If one has a mental model of an ecosystem, restoring the balance of that ecosystem is an appropriate way to treat *C. difficile* colitis. Employing such a mental model, what was initially viewed as a radical approach, namely fecal microbiota transplantation, is a rational therapeutic option. Indeed, clinical trials have now shown cure rates exceeding 90% in patients who have had multiple recurrent *C. difficile* infections.

Increasingly the interactions of a person's behavior and environment with their biologic factors are recognized as contributing disproportionately to the development of disease states. Environmental and behavioral factors affect cellular function and even, through epigenetic changes, the individual's genetic expression and makeup. While medical interventions are important, they arguably contribute less to health than addressing behavioral and environmental issues, especially if the goal is to maintain health or to intervene early in disease states when a return to normal health is possible.

As our students come to understand the disproportionate importance of behavioral and environmental factors on health and the development of disease, it should become apparent that as physicians they will be but one component of a multifaceted "team." Moreover that intervening at the "normal high risk" or "predisease" stages will occur in the community and not in our traditional clinical settings. This will require involvement of community health professionals, Departments of Health, faith-based and fraternal organizations and other groups beyond the traditional medical professionals.

Going forward, students must be provided with a model of health and disease that incorporates the complexities of the relationships among genomic predisposition, the individual's microbiota, and environmental and behavioral factors. Only then the students will develop the mindset necessary to move to specific individualized interventions, or N of 1 treatment approaches, that target the individual patient's problems for optimal outcomes. The first step for medical educators is to adopt such a model and to explicitly reference it as we seek to develop our students' knowledge bases.

Potential learning opportunities to consider:

- Develop laboratory experiences where students sequence the gut or other microbiomes and track changes with interventions such as dietary changes or antibiotic administration to gain firsthand experience and understand the tremendous variability and complexity of microbiomes, how they are affected by environmental changes or medical therapy and how disease states emanate from disruptions in the ecosystem.
- Search databases such as death certificates or Centers for Disease Control (CDC) data and coupled with census track data or other geospatial information explore potential relationships between environmental factors and disease. There are a multitude of data sources that can be accessed to identify environmental factors that potentially affect an individual's health and development of disease.
- Develop simulations of genomic damage secondary to environmental factors such as potential carcinogens allowing the students to study in-silico models.

- Develop case studies or simulations based on genomic variants with well-defined risk for disease development. Have the students research and present interventions that might be considered to either prevent disease development or identify it early in the presymptomatic stage when intervention to restore to health is possible. Examples might include hereditary cancers, QT prolongation and other cardiac diseases, familial hyperlipidemia, and etc.

References

[1] Cooke M, Irby DM, O'Brien BC. Educating physicians: a call for reform of medical school and residency. San Francisco, CA: Jossey-Bass/Carnegie Foundation for the Advancement of Teaching; 2010.
[2] Al-Jashaami LS, DuPont HL. Management of *Clostridium difficile* infection. Gastroenterol Hepatol 2016;12(10):609–16.

Content domains for the 21st century physician

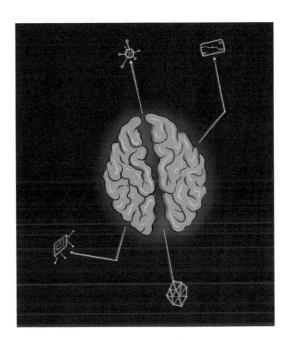

"I think there are four dimensions of science that will require physicians to be differentially trained. The first is in the classic biological understanding. I think that the precision of our understanding of targeted diagnoses, individualized diagnosis, and individualized therapeutics is going to be much finer and far more sophisticated than we ever imagined and someone's going to have to interpret that data. ...I have a hard time figuring out who it is if it's not the physicians... So that dimension of science has to be the physician centerpiece.

<hr /><hr />

Implementing Biomedical Innovations into Health, Education, and Practice
https://doi.org/10.1016/B978-0-12-819620-5.00016-3

The second dimension of science is diagnostic reasoning science.... It's not just complex care but it's really the effortless application of risk assessment and risk targeting, targeted diagnostics and targeted therapy. So this is just part of how we think.

The third is health system science. I think that's a science that is starting to get a little bit of play and starting to get a little bit of place in the curriculum. What's in that box? It's health economics, it's a deep understanding of policy and how political forces interact with each other. It's leadership science. It's understanding motivation and how you engender a team to get them to play well together and move towards a greater goal together. It's quality, it's understanding quality and safety. It's not just being able to do a project, but it's understanding the complexities of root cause analysis and understanding how the simplistic analysis is not going to advance our field forward. It's thinking that there are multiple root causes and we are obligated to tackle multiple root causes for adverse events constantly and we're continuously on this journey.

Then the fourth one I think that's critical for physicians to be facile in is communication science. At the end of the day the patients still are going to rely on us to help them understand the complexities of what they're going through... How do we talk to the patient about their individualized risk, how do we help them understand why we're thinking about the combination of this genetic test and this therapy."

Rajesh Mangrulkar

In Section II we considered the constants in the practice of clinical medicine. While modifications are anticipated, communication, trustworthiness, and ethical and professional grounding will remain foundational aspects of physicianhood. However, envisioned changes in clinical medicine due to biomedical and technologic advances bring with them the need to develop understanding in additional domains and disciplines. Let us consider some of these areas.

Basic technologic literacy

Just as evidence-based medicine principles are used to judge the veracity and applicability of the literature reporting clinical trials and scientific discoveries, medical students and residents will need to become sophisticated consumers of the results produced by artificial intelligence (AI) and machine learning algorithms. While the promise and hype around big data and the power of AI and machine learning is great, the stakes in clinical medicine are also very high. One can ignore an advertisement that erroneously pops up on your computer screen. The consequences are very low for such errors. However, if a decision to treat or not treat is predicated on a flawed algorithm or inaccurate data the potential for harm is real. Hence, just as we educate students to apply the principles of evidence-based medicine to the literature, we need to educate them as to the parallel principles that should govern their use of algorithmic interpretations and recommendations.

Machine learning

As envisioned, the vast quantity of potentially relevant data generated on each individual/patient will not be humanly comprehensible. We will need to rely on computer-generated models and algorithms—machine learning—to aggregate and present the data in an interpretable format. However, as discussed in the machine learning section of this book, there are fundamental principles that should be understood by physicians and hence learned by medical students and residents.

One consideration is that while there are multiple machine learning applications, they are not created equal. To develop a simple analogy, you can use nets, bait casting, spinning or fly-fishing gear to catch fish but there are situations where each method has its advantages. In the machine learning world, neural networks and deep learning networks are powerful tools that have gained attention for their use in image analysis. While the basic approach of neural and deep learning networks are similar, deep learning networks are defined by having more than three layers of nodes that train on a set of features, or data elements, based on the previous layers output. Although they have proven to be powerful tools, the increasing complexity and abstraction means that the data elements used to develop the algorithm are not apparent, the model is a black box. This presents a conundrum for physicians. Can the outcome developed by the deep neural network tool be trusted when the data inputs upon which it is based are unknown?

In contrast, another model is the random forest algorithm. It is a nonlinear supervised learning algorithm that randomly builds multiple decision trees and merges them to arrive at a stable best outcome. One of the advantages of random forest is that the importance of each feature, or data element, is identified allowing decisions to be made regarding dropping features as well as understanding of the algorithm developed.

While it is unnecessary for medical students to develop the proficiency expected of computer science students, they do need to know the kinds of problems for which a certain method is preferred. They should have a basic understanding of how algorithms are generated. For example, how developers use divide and conquer to parse a problem into independent sections with the intention of minimizing compounding error and then aggregating the simple models. How a feature vector consisting of hundreds or thousands of potentially relevant variables is utilized. And the importance of the method used to deal with missing data and the potential implications for the resulting algorithms.

Another consideration is bias in the learning data set resulting in errors. Physicians, residents and medical students need to recognize that commonly used convenience samples and existing databases from clinical care organizations may inadvertently incorporate systematic biases.

A well-known brainteaser told to me by Professor Jagadish nicely illustrates this conundrum.

> A woman is dating two men. She takes a train to visit either of them. Train A takes her to visit man A and train B to visit man B. As both trains run every 10 minutes and she enjoys both of their company equally, she decides to use probability to determine which friend to visit. She decides to simply go to the train station and if train A arrives first board it and visit friend A and if train B arrives first board it and visit friend B. Since both trains run with the same frequency she thought of this solution as a fair way to make the decision. However, as it turns out she visits man A nine times as often as man B, how can that be?

The problem is that the assumption in saying that the odds are equal for train A or train B assumes that the train times are random and they are not. Since the trains run every 10 min, let's say that train A runs on the hour +1 min, +11 min, +21 min, +31 min, +41 min, +51 min. Train B runs on the hour +2 min, +12 min, +22 min, +32 min, +42 min, +52 min. As is evident from this example, if she arrives at the station anytime during the 9 min between the departure of train B and the arrival of train A, train A will be the first to arrive and therefore be boarded. It is only in the 1 min between the departure of train A and the arrival of train B that she could enter the station and have train B be the first to arrive and be boarded. This is an example of systematic bias.

Some clinical data sets are more likely to have such systematic errors than others. A data set from an inner-city charity hospital may be systematically biased by not readily apparent environmental and social determinants of health that are not applicable to a rural population. Or, a database built off of a disease registry may include unrecognized systematic biases that make its relevance questionable to a population that does not have that disease.

Although machine learning algorithms convey a sense of certainty, in reality what is determined is the probability for a given diagnosis or interpretation. There are several decisions that developers must make. It is possible to produce an algorithm that over fits the learning data set rendering it less useful when encountering new inputs. Part of the "art" is determining when an algorithm is sufficiently trained to make it useful in predicting or identifying unknowns.

Another decision is where to set the threshold of probability for making an interpretation. While often the default is to make sensitivity and specificity equivalent, maximizing the area under the receiver operating characteristic (ROC) curve, given the intended use of the algorithm other decisions are sometimes made. If the intention is to use the algorithm for screening the probability threshold may be lowered to enhance sensitivity with the loss of specificity. Conversely, if the intent is to facilitate diagnosis the threshold may be raised to maximize specificity with a loss of sensitivity. These are decisions made by developers but may not be conveyed to the end-user.

While they may not need to understand the mathematical details, medical students and residents should be sufficiently familiar with machine learning to judge the applicability and validity of the outcome of an algorithm and whether it should be applied to their patients. In addition, they should be able to assess whether resulting tools can be relied upon to direct quality and safety initiatives, seen as identifying potentially important new biomarkers or any of the other results forthcoming from machine learning models. While much of this judgment will be based on the congruence of the algorithms outcome with known information, a familiarity with the machine learning algorithmic development process is also important.

Given that machine learning based tools will be ubiquitous in the practice of medicine here is a partial list of considerations to be taught:

- Is the database derived from a patient population representative of that to which it will be applied?
- How was missing data handled? As learning data sets are oftentimes assembled from existing databases, rather than purpose built, there frequently are missing data. In some models, missing variables are replaced by imputation whereby the program assumes a value and inserts it. In others, it is simply absent. The potential problems this might introduce may not be immediately apparent to the user of the output.
- Since the algorithms are developed from historical data how will emerging new diseases or recurring old diseases that are not included in the data be recognized? As supervised machine learning algorithms are based on historical data they cannot predict something that has not occurred in the past.
- How frequently is the database refreshed? What are the curation standards and processes?
- Given the reality that the practice of medicine is constantly evolving, is the output of the algorithm tested on a portion of the initial database that was held out for validation testing or is it tested on a current database? Changes in pharmaceutical administration, sterilization practices, imaging equipment and a multitude of other factors can affect clinical outcomes but not be reflected in databases. It is always preferable to validate an algorithm on a current database.
- How has the problem of bias been considered? All data will have random (stochastic) error, but is there a systematic error or bias in the learning data set?
- Machine learning algorithms provide a probability of a diagnosis or interpretation, not certitude. Developer decisions affect the performance of machine learning models.

Potential learning opportunities to consider:

- Partner with data analysts, computer scientists and experts in machine learning to develop experiments using the big data contained in electronic health records and publicly available databases, such as Medicare and genomic datasets, to teach students not only the potential of but also the assumptions and pitfalls that are inherent in the construction of machine learning algorithms.
- Use multiple different machine learning approaches to derive algorithms from the same database for the same question. Compare the outputs and weighting of features entered into the solution.
- Develop teams, including medical students and/or residents and computer science students, to address current quality and safety concerns. This might be predicting risk of developing a methicillin resistant *Staphylococcus aureus* (MRSA) infection in hospital. Database identification and analysis using different machine learning tools would be the goal for the learning outcome.
- Technology might provide a unique and powerful alternative learning approach. Similar to the example of an early science co-laboratory to support the space science community in their study of the magnetosphere and the reaction of the solar winds, there may be similar "courses" that would allow medical students to electronically observe and even participate in studies seeking to interpret genomic data, identify risk factors for serious antibiotic resistant infections in hospitals, and use Medicare or other databases to identify individuals previously undiagnosed with rare diseases. With thought and creativity, the opportunities are numerous.

An additional benefit of such co-laboratories would be the realization that just as there is risk in the development of machine learning based algorithms by computer scientists who do not understand the realities of clinical medicine, so too there is risk in clinicians utilizing or developing algorithms without understanding the limitations of machine learning.

Sensors

Advances in sensor design and biomedical applications of microelectromechanical systems (MEMS) will result in an array of uses in clinical medicine. The disease status of individuals with conditions such as diabetes, congestive heart failure, Parkinson's, and cancer will be monitored intermittently or continuously in the community, at work and home, allowing timely interventions and therapeutic adjustments. Postoperative patients will be sent home with sensors that will detect signs of infection or other complications requiring early intervention that will obviate the need for rehospitalization. Sensors in the home and on the person will allow elderly individuals to age in place and will decrease the need for institutionalization of people with disabilities.

In clinical settings, sensors will continually monitor patients' vital signs, detect early indicators of sepsis and organ failure, and be integral to the function of machines that will manage anesthesia in operating rooms and ventilator settings in intensive care units. Utilizing sensor data, logistics centers will monitor and oversee inpatients' care in geographically separate clinical units providing sophisticated diagnostic and therapeutic monitoring capabilities for on-site clinicians.

Sensors will be integrated in home diagnostic machines that will make straightforward diagnoses such as viral respiratory infections to more serious conditions recommending professional intervention. Such is the vision for mid-century clinical practice.

What are the implications for the education of medical students and residents? As sensors to monitor health, diagnose and manage acute and chronic disease become ubiquitous the ability to discern true signals from noise or technical error will be important. While a multitude of sensors are currently utilized to diagnose and monitor patients, when unexpected or potentially life-threatening signals are received it is generally easy to verify if it is an accurate or erroneous signal. For example, when an unanticipated EKG pattern is generated the leads are checked to ensure that they have not been reversed or disconnected before a decision to act on the tracing is made. However, such simple measures may not be feasible, given the growing complexity of sensors and their remote deployment. As discussed in Chapter 8, acoustic resonance can result in false signals from accelerometers. In addition to inherent sensor design flaws or unanticipated activation, cybersecurity problems are a growing threat. It is likely that advances in sensor design and "hardening" for other applications, such as autonomous vehicles, will benefit clinical devices as well. That said, medical students and residents need to be taught not to simply accept the output from sensors but to have an appropriate level of awareness when signals might be corrupted. A basic principle is that sensor signals should be verifiable either through direct observation or triangulation with other sensors. As greater sophistication is developed of how best to sort actionable signal from noise in geographically asynchronous real clinical situations additional principles will be forthcoming.

"… There have been enormous advances in signal processing. …Ultimately you'll be able to do some form of triangulation or have redundant readings that you could correlate."

Dan Atkins

A more vexing problem will be the human-sensor interface. This will exist at two levels. The first is working with patients to ensure that they appropriately use/wear their sensor. While it is anticipated that sensor technology and miniaturization will continue to rapidly advance, experience with current sensors suggests that patient engagement will be critical as existing market research shows that physical activity tracking wearables are no longer used by about 1/3 of all purchasers within 6 months. Sensors, just like other interventions, must be used for their value to be realized. To maximize the likelihood that the sensors will be used patients must understand how they will benefit. Well-developed communication skills and the establishment of trustworthy relationships with one's patients will be most important.

The second interface is the vast amount of data produced by sensors that will be available to clinicians. While machine learning generated algorithms hold the promise of helping clinicians by distilling the data into clinically relevant information, the aforementioned caveats regarding machine learning similarly apply to sensor derived data.

Potential learning opportunities to consider

- Build learning experience around examples of sensor data from monitors commonly used in intensive care units. This could include:
  The role of human error—such as ignoring valid alarms, erroneous sensor placement and poorly calibrated instruments.
  The thought processes used to avoid making clinical errors based on faulty data.
  Human machine factors such as alarm fatigue and data overload and actions taken to address these problems.
- Capitalize on the fact that many of our students have fitness tracker wearables. Compare the accuracy of wearable data regarding sleep and physical activity to traditional log data.
- Have groups of students compete to see how sensor data can be manipulated or "hacked" to provide fallacious exercise data.
- Periodically sample students regarding their long-term wearable device use and assess their motivation to change behaviors based on data from the wearables. Use this data as a basis for discussion of how patients might be similar or different.
- Establish collaborations with engineering students and faculty involved in sensor development and deployment to learn the problems they encounter and the concerns they have with different applications as well as the solutions they have developed. The goal is not to turn medical students into technically sophisticated engineers, but rather to instill an appreciation for the potential as well as the pitfalls that sensors provide.

Implementing exercises that highlight some of the technical as well as human variables associated with sensors will provide a basis for understanding the issues they will encounter in their practices.

Material sciences

"...so they need to understand something about material science because the distinction between prosthesis and the original equipment complement is lessening. Increasingly stuff that gets installed in people just stays there a really long time and without which you don't work."

John King

We know that prosthetic heart valves fatigue and fail, that inferior vena cava filters may fracture and result in distant embolization of fractured fragments, that textured breast implants may be associated with the development of an anaplastic large cell lymphoma, and that the materials and manufacturing techniques of prosthetic joints result in different "life expectancies" for the prostheses. The properties of implantable devices, prostheses, and sensors will differ based upon the materials and production processes used in their manufacture.

As the use of implantable devices, scaffolding materials for regenerative medicine, and similar foreign substances are increasingly introduced into patients, an understanding of how the body interacts with those substances will be necessary for at least some disciplines of medicine. Beyond immune reactions which are well known, there is much that is just beginning to be understood. For example: the development and role of biofilms on the surfaces of implants and the implications thereof for the long-term viability of the implant as well as the development of infections; how the body processes foreign materials such as the ingestion of tattoo ink particles by macrophages and transport to lymph nodes, the long-term sequelae of which is unknown; and what will be the fate of various materials used in the 3D printing of implantable tissue scaffolding and devices. In addition to the interactions of the body with the materials, an ever-growing knowledge base will develop of the properties of the materials themselves as they are subject to normal use and stress over time.

It will be important for physicians in training to learn about the properties of materials that are routinely implanted in their patients just as they learn about pharmaceutical compounds. Just as side effects and allergic reactions are discussed as part of pharmacology training, complications of implantables and other devices and foreign substances should be learned as well.

Potential learning opportunities to consider

- Use the FDA's Medical Device Adverse Events (MAUDE) database to develop case studies based on patient adverse events associated with specific materials, length of time device was implanted, or other variables.
- Develop a partnership with colleagues in engineering or dentistry who have material science courses to teach foundational principles and demonstrate properties of commonly implanted materials.

Statistics

"Docs are going to need to be sophisticated about statistics. Docs will need to know what positive predictive value really means, and how one should apply it in a particular situation so that a complicated, expensive, and anxiety-producing search is not initiated for something that didn't deserve that follow-up."

Francis Collins

Whether or not a competency in statistics becomes a premedical requirement, it must be included in the medical school curriculum and reinforced throughout the continuum of medical education and training. The purpose is not to make every physician a statistician but rather to develop knowledgeable consumers of the clinical and scientific literature as well as the reports that increasingly are being generated by "big data."

The ready availability of powerful, easy to use, statistical packages means that physicians will frequently encounter analyses beyond the basic parametric and nonparametric statistics that had been the norm for decades. As we know, there is a difference between statistical significance and clinical significance; physicians need to be more sophisticated in their interpretations of the literature.

How do you interpret the results of a meta-analysis? What are the potential sources of bias that might lead to erroneous results?
When is a multiple comparison correction to avoid a type I error (a false positive finding) needed?
What is the most appropriate statistical test for dealing with data over time?
The questions are many. The answers are central to determining the veracity and applicability of study findings.

To be a knowledgeable consumer of the medical literature requires an understanding of statistics, including what is an appropriate statistical tool for a study, commonly encountered errors, limitations in secondary end point analyses, and the like. To make decisions for one's patients similarly requires an understanding of how to interpret statistical findings. Moreover, as existing statistical methods prove inadequate to deal with the increasingly common massive data sets, new statistical approaches are being developed or modifications made that will require ongoing education of physicians.

"Our residents coming out of medical school are not well-equipped to dissect the medical literature and I think we're not doing a good job teaching them to read the literature and to interpret it.

There is a piece in JAMA Cardiology about a network meta-analysis showing that the optimal blood pressure to treat patients to is between 120 and 124 and there's a very big reduction in the hazard rate at those levels compared to 124-130, 130-134, etc.

If you were to ask your graduating students, 'What is a network meta-analysis?' they're going to be clueless.

I savaged this paper with the media. I said, 'Meta-analysis is not a reliable method for integration of knowledge and here are all the flaws.' In this type of meta-analysis you don't actually directly compare things, you compare one thing to another and then a third thing and then you input what the actual relationship would be. That's what people don't understand. They don't understand how a network meta-analysis can be wrong.

If you're an internist and you read that and you start treating your elderly patients to that level, you're going to have fractures and falls and subdural hematomas and everything else. I said, 'This is not good science,' and that's what you need to know to be a great doctor, we have to teach people.

How well are we doing at positioning people to be appropriate doctors? Which means to understand what's solid and evidence-based, what's speculative, what's hype, what's not hype. That's why I emphasize these issues in training to interpret the literature.

Learn contemporary statistics. There's no substitute for understanding this stuff."

Steven Nissen

A working knowledge of the appropriate use and interpretation of statistics is requisite for a professional conversant in the language of medicine and science. To ensure statistical literacy, the appropriate application of statistical methods should not be simply a didactic course. Rather, this material should be an integral part of any discussion or learning exercise built on scientific and clinical evidence. It should not be simply assumed that medical students and residents will develop a robust understanding of statistics or that a course will suffice. Statistical literacy is as fundamental as learning anatomy to the development of a future physician. Medical educators need to ensure that it occurs.

Potential learning opportunities to consider

- Evidence-based medicine curricula where understanding the use of statistical analyses as reported in the literature is key to interpreting studies. Including examples of the misuse of statistical tests can enhance the value of the learning experience.
- As part of their genetics curriculum, have students compare anonymized "patient" genomic results with publicly available population databases to determine the potential implications for disease of an observed variant. Developing an appreciation for the importance of a representative control group and how findings are expressed with the goal of establishing a basic understanding of the statistical and association analyses that go into such data interpretations [1].

Health system science

The ultimate goal of medical education is the delivery of clinical care. While this is self-evident, formal medical education programs historically allocated little, if any, time to equipping their learners in the provision of care. Health systems science is characterized as the third medical science joining the basic medical and clinical sciences. It is defined as "the principles, methods, and practices of improving quality, outcomes, and costs of healthcare delivery for patients and populations within systems of medical care" [2]. Topics included in health system science range from health policy to the structure and function of clinical delivery systems to informatics. The defining characteristic is a holistic view of factors that must be accounted for when considering how to improve health and care delivery at the individual and/or population level. Let us look at some components of health system science.

Systems engineering

"I think systems engineering work can be transforming to the delivery systems of care to better prepare them for the customer requirements. It could be in terms of a brain tumor or dialysis or primary care. I think that we ought to promote systems in the curriculum."

Gary Kaplan

Reductionism has been a dominant force in biomedical research as has subspecialization in clinical medicine. Systems engineering is just the opposite and is a central component of health systems science. Embracing multiple disciplines within engineering, components of organizational psychology and project management, it looks at the totality of a problem or process ranging from the technical to the human factors involved.

As the provision of clinical care is a complex activity, a holistic approach, as embodied in systems engineering, is required to optimize the process. The mindset and tools of systems engineering can be applied to a wide variety of issues encountered in biomedical science and clinical practice. Problems ranging from clinical wait times to "noncompliance" of pregnant women with prenatal visits to nosocomial infection outbreaks, all can be addressed using the principles and processes of systems engineering [3].

A common application is addressing clinical quality and safety problems. It begins by a thorough analysis of what the real problems are that need to be resolved and provides a logical process to determine root causes as well as potential options (RCA2—Root Cause Analysis and Action [4]) to make improvements or resolve the problem. Often times pictorial representations of the processes and relationships are used to capture the complexities, multiple stakeholders and decision points involved. Techniques that are employed in process improvement such as flowcharts, Pareto charts, and fishbone diagrams are among the useful tools with which learners should be familiar. While introduction to the terms and techniques is possible through didactic presentations, involving learners in their application through case studies, simulations or participation in groups wrestling with real problems is preferable.

For medical student and residents, developing an understanding of the thought processes and logic underlying systems engineering will equip them with a valuable skill with widespread applicability in their professional careers. As with many aspects of medicine, this requires not only knowledge of the tools and processes to use, but also experience in their best application.

Potential learning opportunities to consider

- Involve medical students and residents as members of the team addressing quality and safety problems. Being immersed in projects that apply the systems engineering concepts of analysis to deeply understand a problem before identifying solutions, then devising and implementing solutions can be a powerful learning experience.
- Develop case studies based upon real quality and safety problems addressed in your institution. Have students work through the problem definition, scope of inquiry, steps involved in the root cause analysis, and the different components of the delivery system that are

involved in rectifying the problem. Then follow up with the involved units to learn how the implementation of recommendations was accomplished and subsequent performance changed.

- Have students identify and address clinical problems. An example is elevated lead levels in children. The impairment of intellectual development and other effects are well-known. Physicians know how to screen for lead toxicity and how best to treat it, but based on the health/disease continuum model, identifying normal children at high risk and focusing screening and interventions to prevent lead ingestion on these children and their families is a better approach. To do so requires understanding the potential sources of lead, resources such as census track data that includes housing information, and community and governmental resources that can be enlisted to educate the parents regarding prevention and abatement options. Not only would such a learning opportunity demonstrate the power of the systems-based approach, it would introduce them to the multitude of information sources and expertise outside of the clinical delivery system that can be enlisted to address important clinical concerns.

Leadership, management, and Teamship

"Our trainees are instilled with excellent medical IQ, but we do little to help them build their leadership EQ. Medicine is a team sport, and our next generation of clinical leaders must learn how to work collaboratively—with the patient at the center—to manage the inevitable dissenting opinions that will arise when you have a group of smart, driven people work together. AMCs need to make formal leadership training a required part of the curriculum."

Elizabeth Nabel

Although physicians have long been thought of as leaders in their practices, the hospital setting, and their communities, the importance of formal attention to leadership in medical education is a relatively recent development. But a conundrum for medical educators is the fundamental question "What is leadership?" We use the terms leader and leadership in connection with the surgeon performing a major operation supported by a team of professionals including anesthesiologists, nurses and technicians; the chair of a clinical department; the oncologist working with a team of professionals caring for patients with breast cancer; and numerous other examples.

Popular notions such as "everyone is a leader," while true in some respects, further exacerbate the confusion. Interestingly, however, rarely are management skills highlighted as important for future professionals.

There is a perception that management skills are less important than leadership skills. Perhaps it is because the intrinsic value many people associate with a management role is deemed less than a leadership position. That said, the responsibilities and characteristics of managers and leaders are often conflated and successful leaders spend much of their time and effort on management.

Recognizing that clarity as to what is meant by leadership responsibilities as distinct from those of managers is not forthcoming, for our purposes I will use John Kotter's definition: "Management is a set of well-known processes, like planning, budgeting, structuring jobs, staffing jobs, measuring performance and problem solving, which help an organization to predictably do what it knows how to do well. Leadership is about vision, about people buying in, about empowerment and, most of all, about producing useful change. It is associated with taking an organization into the future" [5]. While this definition is focused on the business world, it readily transfers to the clinical world. I recognize that there is overlap and that a clear distinction between management and leadership is somewhat arbitrary.

In the context of Kotter's definition, it is apparent that the success of many clinical activities depends upon good management. Performing an operation, providing high quality coordinated care to cancer patients, and a multitude of clinical activities where the processes and steps needed for optimal patient outcomes are defined require good management and managers. The importance of management must not be underestimated and the value of good managers cannot be minimized. The lack of attention to the "business side or management" of practice has been recognized as a recurring failure on the part of medical educators for decades. Given the direction clinical medicine is heading, it is likely that it will become even more important as physicians will be part of and charged with managing care teams that include an array of professionals and technical people.

Medical students and residents must know how to use the tools and skills of good managers as that will be foundational for their professional success. What are some of those beyond the processes Kotter enumerates, many of which are technical skills? Managers need to be able to ask for and receive feedback and suggestions without becoming defensive or belittling the input. They must understand team dynamics: how to select, support, and develop people; how to motivate individuals; how to delegate responsibility; how to manage conflict; and how to ensure group and individual accountability. Skills essential for managers and leaders include communication, building and maintaining trust and very importantly maintaining their humanity—being able to understand the concerns and perspectives of others—emotional intelligence.

What are the unique characteristics of leadership? Leadership is focused on constructive change. It requires being aware of external and internal factors that will impact your group's or organization's ability to provide care, identifying new opportunities and challenges to existing operations. Then developing and communicating a vision as to what needs to change, why change is needed, how it will affect individuals, and the path forward. This requires creativity, a proclivity for taking risks, a willingness to make decisions often in the face of uncertainty, and the ability to convince others that the vision, albeit often risky and requiring uncomfortable changes, is worthy of their participation. Indeed, the adage "leaders need followers," while self-evident, is too often an afterthought and failure is the result. But beyond a well communicated vision, the leader's role is to ensure that the resources and tools necessary to accomplish the vision are secured. While perhaps less obvious than the "vision piece" this is critical.

We all can recall articulate and compelling visions for change, often times forthcoming from politicians, devoid of any detail as to how it will be accomplished. People line up to follow the individual but when no tangible results are visible cynicism results. Working to ensure availability of the resources and infrastructure necessary to accomplish the vision is the essential but less visible nitty-gritty work of leaders. As clinical practice is changing at an ever accelerating pace, for many physicians, leadership skills will be as necessary as management skills. Being a leader and exerting leadership is not dependent upon a title but reflects action, responsibilities, and attitude.

Given its importance, what approach should be taken to develop medical students' and residents' leadership capabilities? As business and industry has recognized the importance of leaders for many decades, leadership curricula have primarily been the province of business schools. While many leadership models have been developed, essentially all share common components.

Building upon existing models, as shown in Fig. 16.1, at the center is personal understanding. Being self-aware and knowing one's core values is essential. Every leader engages in multiple levels of interaction ranging from interpersonal dyadic relations to team-based to institution/organization level relationships. Broadly speaking, there are four domains of leadership—character, context, communication and competence. In each domain the level of interaction often times requires different approaches, different skill sets and knowledge. Understanding the differences and nuances that the interplay between level and domain create is central to effective leadership [6]. For example, in the domain of communication, interpersonal interactions require leaders to develop the ability to engage in effective conversation around difficult topics. At the institution/organization level to convey a compelling vision. Depending upon the situation, leaders are also members of teams. Effective leaders role model teamship and easily transition from leader to follower as conditions dictate.

LEADERSHIP / TEAMSHIP

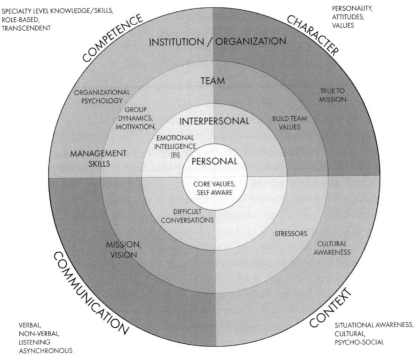

FIG. 16.1 A framework of leadership/teamship depicting the interactions between domains, level of interactions and applicable understandings and actions.

As the specific leadership roles that medical students and residents will fill in their professional careers are unknown, the emphasis should be on broadly applicable knowledge and skills. As depicted in this framework, and common across leadership models, is the importance of knowing oneself and how we are perceived by others. There are multiple self-assessment tools such as 360 feedback that may be utilized to assist individuals develop self-awareness and understand their strengths as well as what is seen as detracting from their effectiveness. There are certain characteristics that appear to be important regardless of role or situation such as self-confidence, integrity, humility, and optimism.

While competence will be more specific to the particular roles and responsibilities, management skills, an understanding of motivation, and emotional intelligence transcend most situations.

Given the rapid changes foreseen in clinical medicine, an appreciation of the context, or situational awareness, in which the leader is operating will be increasingly important. This includes the organization's culture, the stress that people in the organization are under and how critical the situation is. For trainees, this domain of leadership may seem ambiguous. Case studies and simulations may be useful educational tools.

The last domain is communication. It is essential not only for leadership, but virtually every activity physicians will engage in. As was discussed in Chapter 4 in the section on communication, listening skills are essential. This is especially true in leadership roles.

"Everything that we do in medicine is being part of an effective team or leading an effective team, and it is so easy for us to get angry at something that's not working inside the hospital, inside your office, in whatever, and just say, 'Oh, they're not doing their job, they're lazy, they're unmotivated,' and get frustrated by that and not understand why that team isn't performing to its ability and your expectations.

I think team dynamics and understanding how a team works, how to be a team member, how to lead a team, and how to properly assess and motivate a team, that's really hard. If you can't assess them properly, you're not going to be able to motivate properly. You have to deeply understand how teams function and individuals operate and assess their ability.

You know your job as a leader is to provide opportunity and to motivate capability, and before you can motivate capability you've got to understand what capability is. I think that everything we do to try to be better involves a team around us. I think it's really central to success."

Stephen Papadopoulos

As it is expected that most physicians will work in large groups and be members of care teams that may include individuals outside of their employment institution, an understanding of the complexity and nature of organizations will be required. Different models of governance, management and administration, approaches to strategy, planning, and budgeting exist in different organizations and even within an organization in different units. Conscious consideration of these differences when making decisions will facilitate team function and will decrease misunderstandings that may occur as a result of unrecognized competing priorities and differing incentives.

In some clinical care settings, teams consisting of individuals from different professions have long been recognized as necessary to provide care. Cardiac and transplant surgery teams are readily identified examples. From a systems approach, the vast majority of care, albeit sometimes not recognized, involves teams. Thus, understanding team dynamics and function is essential to the education of future physicians.

While leadership and management are familiar terms that have been extensively studied, written about and taught for decades, the concept of teamship is less developed and some would argue simply a neologism for what constitutes a high functioning team. For our purposes, teamship refers to the characteristics and functions of the team. It is often expressed in relationship to athletic teams but is relevant to everything from entrepreneurial startups to clinical care teams.

Distilling findings from studies of teams in domains other than clinical medicine, commonly cited key factors for "high-performing teams" are shared values, strong mutually respectful relationships, a shared identity and understanding of each individual's role, and a willingness to sacrifice individual goals and recognition for the sake of collective success. In 1965 Tuckman published the oft referenced four stages of team development—forming, storming, norming, and performing [7]. The initial stage of forming requires thought as to what skills and perspectives are needed. Once the team members are recruited, the work of developing a shared vision and approach to accomplishing the intended purpose of the team begins. This is the storming stage. Often times team members do not know one another, there is disagreement as to the best approach to the task, and members have different communication and interaction styles. As familiarity develops, explicit or implicit group norms develop, a shared vision for how the task will be accomplished emerges and work toward the goal begins to progress. If these stages have been successfully navigated, a successful well-functioning team has developed. As expressed by Thomas Lee during the New England Journal of Medicine Catalyst webinar on July 25, 2018, these cohesive high functioning teams have four characteristics that he termed "grit."

- First, there needs to be a goal that is meaningful to every member of the team. Oftentimes there is a hierarchy to the goal that connects the menial everyday tasks to an idealistic goal. The example used was a goal of reducing suffering of patients. Tasks such as filling prescriptions, answering emails, making rounds and similar activities speak to coordinating care, all in the name of the overall goal of reducing suffering.
- The second characteristic is a growth mindset. This is the belief that change can happen and as a team they can get better.
- The third characteristic is resilience. Setbacks will be encountered. The ability to recover from the setbacks and the flexibility to deal with unexpected occurrences is a characteristic of highly functioning teams.
- The fourth characteristic is identification with the group. It's pride in being part of the team and a member of a group that is dedicated to achieving something greater than can be accomplished individually.

But when clinical teams are studied these are convenient oversimplifications of complex relationships. One problem is that the complexity of clinical care is such that clinicians involved in one facet of the patient's care may be only vaguely aware of the importance and contributions of professionals involved in other aspects of the patient's care. What are the boundaries that define a clinical team? A cardiac cath team or a surgical team is readily defined as it is bounded by the four walls of the cath lab or operating room. But taking a systems approach, social support agencies, faith-based communities, family and a variety of others all may play important roles in a patient's care but not be readily apparent. Acknowledging these realities, educational programs on teamship should focus on identifiable clinical teams that include multiple professions occupying different positions in the traditional hierarchy, are not readily defined by a location, but rather by caring for a group of patients or addressing a function or problem.

To participate in and lead team-based care requires a multi-perspective understanding of the complexities involved. As has been repeatedly demonstrated in athletics, through quality teamship, a less talented and seemingly weaker team when looked at individually can triumph over an individually more talented team. Quality clinical care requires competent individuals, but that is not sufficient, it also requires competent teams—termed collective competence. Not surprisingly, studies of teams in acute care contexts show that akin to good teams in other domains, information sharing, alignment of goals, clear responsibilities, and mutual trust are important features of high functioning acute care teams. However, all who work in acute care settings know that differences in problem framing, solutions and prioritization based upon professional or specialty perspectives are a daily occurrence [8]. Does this imply that even though different disciplines are caring for the same patient they are not functioning as a team? Or, does it require us to be nuanced in our thinking about teams and the characteristics that make them successful?

Indeed, the results of a study, primarily focused on ambulatory care, of distributed teams caring for patients with advanced heart failure presents some interesting observations. Through 183 interviews with patients with advanced heart failure and professionals from a wide range of disciplines, relationships between team members' goals, understandings, and values, as well as the collective competence of the team were explored. Different team members did not always share an understanding of patients' needs and concerns, future care plans such as homecare support, or patient's wishes regarding death and dying. Surprisingly, the effect of this misalignment was variable. Sometimes agreement and other times disagreement appeared to support the team's collective performance [9]. This departure from the usual view of healthcare teamwork requiring mutual assumptions and coordination raises questions as to its importance. Are there major differences in teams based on whether they are primarily ambulatory or acute care inpatient focused, do differing methodologies account for findings or have studies simply not recognized that even lacking completely aligned understanding, goals and vision, "good" care can be delivered?

As many clinical functions, ranging from quality and safety improvement to patient care, involve teams, teamship arguably is as, or more, important than traditional management and leadership skills. Moreover, as many teams in clinical settings are comprised of "knowledge workers" who are essentially independent contractors, understanding how to mold a high functioning team is important. What then should we strive to teach our medical students and residents about teamship? Here are a few suggestions:

- Unique to the concept of teamship is how participants are enlisted and their support requested and received. Attention to clearly communicating the "why" an action is being considered or taken is important to enlist buy-in and lay the foundation for a cohesive highly functioning team.
- Thoughtfully consider the diversity of team members required to ensure that skill or knowledge gaps do not compromise function.
- Realize that members of the team will frame problems differently and may even differ on what is thought to be a problem.
- Cultural differences based upon professional and discipline socialization will be present.
- Issues of professional hierarchy, even if unspoken, will be present.

While the importance of teams is unquestioned, we still have much to learn about their optimal configuration, how best to structure them for given functions or tasks, and which leadership and teamship qualities of their members are most likely to result in high functioning teams. As research provides greater understanding, medical educators will need to incorporate this knowledge into their educational offerings.

Potential learning opportunities to consider

Leadership, management and teamship skills cannot be developed through rote learning. Providing opportunities for students to acquire and refine these skills is necessary.

- Obviously extracurricular activities, student government, and roles in external student organizations whether at the national or regional level provide opportunities for students to develop management and leadership skills. Couple this with reflection and/ or debriefing to assist the student in identifying their strengths and weaknesses and suggestions as how best to enhance skills needing development.
- Develop leadership electives for students interested in furthering their skills. Invite them to observe institutional leadership meetings. Then have the students reflect on what they observed, tying it to their knowledge of organizational psychology and leadership, and discuss their observations and reflections with a skilled leadership coach.

- In courses involving small group work, explicitly rotate "leadership" with expectations that all group members will serve as leaders. By incorporating self and peer feedback not only will the leader benefit but other members of the group will be engaged in understanding what enhances team functioning.

Deskilling

One consideration is deskilling. A valid concern is the dependence that the introduction of powerful machine learning generated algorithms will have on physicians and other clinical personnel. Deskilling is likely to occur [10]. There will be an ever-decreasing number of physicians with the training and extensive experience necessary to accurately interpret the raw data. If algorithmic interpretation of essential diagnostic and therapeutic monitoring data becomes the norm, how will clinicians develop the skills needed to confirm the veracity of the machine generated results? An example that you may have experienced is the inability of many checkout clerks to calculate change if their electronic cash register isn't working. If you desire to rid yourself of extra change and give the cashier $23.14 for a purchase of $22.89 you are likely to get a puzzled look or even be told that you only need to give them the $23.00.

A conundrum that medical educators will need to resolve is what should be learned about interpretation of primary data when computer-generated reports will exist and become the standard for images, electrocardiograms, electroencephalograms, sleep studies and all such digitized data. Should the emphasis be on a general knowledge of the technology, how to "read" findings that have major clinical significance and risk such as acute MI, or should primary data interpretation skills be viewed as of historical significance only and deserving of the same fate as how to use a flame photometer when automated analyses of sodium and potassium became readily available? While this decision will be important for medical student curricula, it has major implications for graduate medical education.

What to eliminate

As the period of education and training is not infinite, consideration needs to be given to what should be eliminated. Long-standing cherished traditions in medicine are rapidly changing. For example, what is the evidence for many of the aspects of the physical examination? Should students continue to be taught how to take a blood pressure using a sphygmomanometer and auscultation or rely on an automated blood pressure monitor? Just as CT scans and MRIs have supplanted a lengthy and sophisticated neurologic examination in the localization of central nervous system lesions it is likely that technology will render obsolete other physical examination skills. For example, machine learning derived algorithms may well be coupled with real-time auditory and ultrasonographic cardiac images to assess cardiac function. Should the focus be on teaching medical students ultrasonography rather than cardiac auscultation?

New technology should not be used simply to augment traditional methods of instruction but rather to totally replace them. An obvious example is gross anatomy. While the hours devoted to gross anatomy and dissection have been curtailed in most medical schools, if the learning goal is to develop an appreciation for anatomic relationships, currently available and evermore sophisticated anatomical learning programs coupled with plastinated specimens should supplant gross dissection. Rather than early in all students' medical education, dissection should only be part of the educational program for students committed to careers in selected specialties such as the surgical disciplines.

While it is anticipated that growth in understanding of pathways leading to disease will be incorporated into the educational program, should there be a concomitant deletion of material that teaches disease classifications based upon morphology? Just as it is now recognized that cell surface markers are necessary for accurate diagnosis of hematologic malignancies, it is expected that there will be similar biomarker and genetic classifications of other tumors. Should histology be eliminated other than as an interesting example of how pathophysiologic disease associated changes result in visible cellular changes in diseased as compared to normal tissues?

Should discipline based clinical rotations be replaced by longitudinal involvement and learning from patients with specific prototypical diseases?

These are but a few of the potential opportunities to eliminate existing curricular offerings to free up education time. All such decisions will require a thoughtful assessment of what is being lost and whether the perceived gain of what is being added justifies eliminating current content. Deciding to eliminate or markedly curtail existing curriculum will be difficult. As such, it is worthwhile to recall a few earlier stated principles:

Focus on the future professional. We must resist the desire to create the optimal curriculum and learning experiences for the current practice of clinical medicine.

Focus on learning. Faculty are oftentimes conflicted as their desires and needs no longer align with what is best for the medical students and residents. Service needs, faculty reluctance to learn new skills such as coaching rather than lecturing, and financial arrangements may conflict with optimal learning situations.

Most importantly, Teach how to think, not what to think. Given the foreseen growth in our understanding of health and disease and the proliferation of tools and options for physicians, the focus must be on how to think. How to reason. How to frame problems and options. Not on what is currently thought to be the right answer or approach.

References

[1] Richards S, Aziz N, Bale S, Bick D, Das S, Gastier-Foster J, et al. On behalf of the ACMG laboratory quality assurance committee. Standards and guidelines for the interpretation of sequence variants: a joint consensus recommendation of the American College of Medical Genetics and Genomics and the Association for Molecular Pathology. Genet Med 2015;17(5):405–24.

[2] Skochelak SE, Hawkins RE, Lawson LE, Starr SR, Borkan JM, Gonzalo JD, editors. Health systems science. Philadelphia, PA: Elsevier; 2017. p. 11.

[3] Swensen SJ, Dilling JA, Harper CM, Noseworthy JH. The Mayo Clinic Value Creation System. Am J Med Qual 2012;27(1):58–65.

[4] DeRosier J. RCA2 improving root cause analyses and actions to prevent harm. National Patient Safety Foundation; 2015. http://www.ihi.org/resources/Pages/Tools/RCA2-Improving-Root-Cause-Analyses-and-Actions-to-Prevent-Harm.aspx.

[5] Kotter JP. Management is (still) not leadership. Harvard Business Review 2013. Digital article, https://hbr.org/2013/01/management-is-still-not-leadership.html?utm_source=feedburner&utm_medium=feed&utm_campaign=Feed%3A+harvardbusiness+%28HBR.org%29.

[6] Callahan C, Grunberg NE. Military medical leadership. In: Schoomaker E, Smith D, O'Connor F, editors. Fundamentals of military medical practice. Washington, DC: Borden; 2018.

[7] Tuckman BW. Developmental sequence in small groups. Psychol Bull 1965;63(6):384–99. https://doi.org/10.1037/h0022100. PMID:14314073.

[8] Lingard L, McDougall A, Levstik M, Chandok N, Spafford M, Schryer C. Representing complexity well: a story about teamwork, with implications for how we teach collaboration. Med Educ 2012;46(9):869–77.

[9] Lingard L, Sue-Chue-Lum C, Tait GR, Bates J, Shadd J, Schulz V. Pulling together and pulling apart: influences of convergence and divergence on distributed healthcare teams. Adv Health Sci Educ 2017;22:1085–99.

[10] Cabitza F, Rasoini R, Gensini GF. Unintended consequences of machine learning in medicine. JAMA 2017;318(6):517–8. https://doi.org/10.1001/jama.2017.7797.

Additional resources

- Silver N. The signal and the noise why so many predictions fail—but some don't. New York, NY: The Penguin Press; 2012.

- Pink DH. Drive the surprising truth about what motivates us. New York, NY: The Penguin Group; 2009.

- Senge PM, Kleiner A, Roberts C, Ross RB, Smith BJ. The fifth discipline field book. Strategies and tools for building a learning organization. New York, NY: Crown Publishing Group; 2014.

- Skochelak SE, Hawkins RE, Lawson LE, Starr SR, Borkan JM, Gonzalo JD, editors. Health systems science. Philadelphia, PA: Elsevier; 2017.

Graduate medical education

The coordination between medical school and graduate medical educa-tion programs will likely increase and the competency level required for graduation from medical school will be in large part determined by what is expected at the beginning of a resident's graduate medical education program. Depending upon the direction they envision as the next step in their education and training journey, medical students will graduate with variable levels of demonstrated competency. For example, the emphases and level of proficiency for communication competencies will differ for a student planning on pursuing a procedural discipline as compared to a mental health discipline. The skills associated with informed consent are expected to be developed to a higher level for the former student whereas skills such as motivational interviewing and active listening for the latter

Implementing Biomedical Innovations into Health, Education, and Practice
https://doi.org/10.1016/B978-0-12-819620-5.00017-5

student. For this to happen, the artificial schism between "undergraduate" and "graduate" medical education must be bridged. Agreement not only on competencies but also on level of attainment and the assessment methods utilized to demonstrate achievement can only be reached through intense discussion and development at a national level.

Specialty obsolescence and emergence

Readily apparent from the vision of clinical practice set forth in this book is the likelihood that existing specialties will shrink dramatically or even disappear. How will specialists such as radiologists who potentially will be replaced by computers respond? Will there be impediments to implementation of sophisticated and accurate machine learning based algorithms for the interpretation of digitized data? Will professional organizations seek through reimbursement, political and regulatory means to delay or block their adoption for image interpretation? The existential questions raised by scientific and technologic advances will be major disruptors and the transition from current state to future state will be tumultuous as it is human nature to protect one's professional status and livelihood.

From a graduate medical education perspective there are multiple concerns. Will disciplines at risk of significant disruption from technologic advances seek to redefine themselves with major implications for curricula and learning experiences regardless of their faculty members' expertise? Examples are pathology and radiology that have discussed steps to redefine themselves as the diagnostic disciplines. However, for decades clinicians in these disciplines have had minimal direct patient contact and their faculties likely do not have the deep diagnostic acumen beyond image and laboratory study interpretation to equip residents with a robust array of diagnostic skills. Will collaborative partnerships arise with faculties from other specialties with direct patient contact and diagnostic expertise? Another example is oncology. The potential of liquid biopsies being used to diagnose, define the tumor's pathophysiologic pathway and therapeutic sensitivity, akin to the widely used culture and sensitivity for bacterial infections, begs the question as to the need for subspecialty training in oncology. For trainees and for medicine at large, would it be better to markedly shrink the number of training programs in such disciplines or cease training altogether?

"Let's not forget the policy, regulation, funding and politics around education. The federal government through Medicare is the primary source of financing for medical residency training (a.k.a., graduate medical education (GME)).These payments shape GME to a large extent by determining the number of trainees, geographic allocation of residencies, and more. Furthermore accrediting bodies, such as the Accreditation Council for Graduate Medical Education (ACGME), oversee the standards and content of GME training.

If we don't pay attention to these factors, we're overlooking some of the key determinants of GME. Indeed, there have been efforts and proposals to reform the GME payment and policies, without much progress. Since it will take an Act of Congress, to change the GME payment system, politics around this issue can complicate efforts to enact change."

Victor Dzau

While there are umbrella organizations for programmatic accreditation and certification bodies, I am pessimistic that decisions regarding the elimination of specialties and their attendant graduate medical education training programs will be made on the basis of societal needs and the realities of clinical medicine. Rather, political and economic constraints will result in an uncoordinated shrinkage of programs and potentially chaotic changes in scope of practice among specialties. Residents may well find themselves trained for specialties made obsolescent by technologic advances. We owe it to our trainees and future colleagues to ensure this does not happen.

Not only will scientific breakthroughs and technologic advances fundamentally disrupt clinical practice in midcentury, or even render some specialties irrelevant, these breakthroughs and advances will also create a need for new expertise. Existing specialties will embrace new areas of expertise and expand graduate medical education training to develop the requisite knowledge and skills in their residents. One can envision that new disciplines with epistemological roots in engineering and biomedicine or sociology and medicine will emerge. Given the likely pace of disruption, ensuring that emerging disciplines and their graduate medical education programs are developed in a thoughtful manner will be a challenge.

Change in graduate medical education

Of immediate concern is the conflict between service and education. As discussed earlier, reimbursement models, the tyranny of efficiency, the reliance upon residents to provide clinical care, and regulatory requirements have had unintended deleterious effects on residents' education. But it is important to recognize that academic clinical faculty are in part responsible for some of these problems.

Breaking the cycle and shifting the primary focus to learning in the context of providing care rather than providing clinical care in the context of an academic medical center that also educates will require leadership and support from the faculty, the care system administration, payers, regulators and, not to be forgotten, the patients themselves. While strides have been made in this direction through administrative decisions such as limiting the size of services, duty hours, night floats, and nonteaching services the basic problem remains that educational considerations are too frequently secondary to clinical care—service—demands.

Graduate medical education (GME) should be the time in a medical professional's development when their mental models of health, disease—its presentation and natural progression, and approaches to intervention are essentially completed. To do so requires that residents will encounter the full spectrum of patients and disease stages likely to be cared for in their clinical practices. This will likely require experiences outside of the academic medical center hospital and clinics. Importantly, not only are the patients necessary, but so too are the "faculty" who will provide guidance and feedback. Rather than the current prevalent model of "learning from whatever walks in the door" I advocate for an intentional effort to ensure the sophisticated, broad-based knowledge and experiential foundation of every resident.

Modifications to existing educational experiences are required in many programs. For example, if residents are to develop the skills necessary to function in and lead teams of professionals, suitable immersive training experiences must be developed. For some situations, such as surgical teams and trauma teams, these opportunities are integral to specialty training. However, for many specialties, especially when the care teams include professionals outside of the training institution, intentional education experiences must be developed. This will potentially require the teaching faculty to also expand their concept of the care team. For example, as reimbursement models and quality indicators evolve to functional outcome for joint replacements, essential members of the care team will include many other members in addition to the readily recognized surgical and floor nurses and the orthopedic surgeons. The physical therapists who provide preoperative physical conditioning and postoperative rehabilitation, the pain management team who facilitate early rehabilitation by minimizing the patient's discomfort, the home healthcare service that ensures appropriate medical equipment is in the home allowing early transition of the patient out of the hospital or rehabilitation facility, and the telehealth support staff that facilitate virtual follow-up of the patient after discharge will all be part of the care team.

It is likely that all GME programs will require some degree of new knowledge and skill development. As previously mentioned, expanded training in biostatistics; enhanced communication skills; applications and limitations of sensors; artificial intelligence including machine learning algorithms; robotics; and more will require the specific attention of new or augmented curricula and learning opportunities. For graduate medical education programs, learning will need to be tailored to meet the expected clinical needs of their graduates, building upon the more general knowledge base developed during their residents' medical school education.

For all GME programs, it will be relatively easy to identify the immediate needs but more difficult to recognize implications of future scientific discoveries and technologic advances. As such, an important task for faculty involved in GME will be developing in their residents the skills and aptitudes necessary to ensure that they are equipped to continuously learn and reinvent themselves throughout their professional career as clinical practice patterns are disrupted and/or evolve. Doing this will require creativity and rigorous study as initiatives such as self-assessment to identify knowledge and skill gaps coupled with lifelong learning have been discussed for decades. However, empirical studies of physicians in training and practice have been disappointing. The recurring result is that self-assessment is poorly correlated with objective findings. Perhaps this is a universal attribute of humans that is not amenable to education or training. Rather than accepting that self-assessment is invariably faulty, concerted effort and rigorous study is needed to discover and implement effective educational interventions. Arguably, accurate self-assessment coupled with the skills to continuously learn and "reinvent" themselves will be among the most important aptitudes for faculty to instill in their GME program graduates.

Reflection

What are the implications of technologic and scientific advances for my specialty? Will it remain relevant in essentially its current form or need dramatic change?
Should residency training in my specialty include learning experiences in engineering, public health, or other disciplines outside of traditional clinical medicine?

Additional resources

- Wachter R. The digital doctor: hope, hype, and harm at the dawn of medicine's computer age. New York, NY: McGraw-Hill Education; 2015.
- National Academies of Sciences, Engineering, and Medicine. Graduate medical education outcomes and metrics: proceedings of a workshop. Washington, DC: The National Academies Press; 2018. https://doi.org/10.17226/25003.

Educational directions and challenges

Not only must we consider what should remain and change in the content of medical education due to advances in our biomedical knowledge
base and technology, so too we must consider what should remain or
change due to developments in educational process and clinical practice.
Competency-based medical education and interprofessional education
are two examples of educational process changes that were recommended
in the 2010 report *Educating Physicians: A Call for Reform of Medical School*

Implementing Biomedical Innovations into Health, Education, and Practice
https://doi.org/10.1016/B978-0-12-819620-5.00018-7

and Residency [1]. But there are also educational practices and venues that deserve strengthening due to their value as educational tools. In this chapter we will consider innovations and traditions deserving of a place in our medical education programs.

Competency-based medical education

All involved in medical education understand that the learning journey followed by each medical student and resident is unique. Learners vary in their abilities to understand material and to develop proficiency in procedural skills. Whereas the public expects that physicians all possess a high level of knowledge, skill, and aptitudes, medical student and resident advancement continues to be based on time spent on task, provided that a minimal standard is met, rather than on demonstrated expertise. For the past century, the proxies for achieving competence have included graduation from an accredited medical school and completion of a residency training program, national single point in time examinations such as the United States Medical Licensure Examination (USMLE) and board certification examinations, and state licensure. However, many studies have documented the variability in medical school graduates' knowledge and abilities, in the performance of residents completing their graduate medical education training programs, and most concerning to the public, in practice patterns and patient outcomes.

In recent years, there has been growing interest in competency-based as opposed to the traditional time-based education [2], representing a shift in the perspective of medical educators. Taking this a step further, the idea that a learner would move through the educational process based on their demonstrated achievements independent of time on task is intriguing. Not only could this address the issue of variability in competence, but also for many medical students and residents it might shorten the length of training. As promising as the potential of time-independent competency-based medical education appears, there are important considerations. Let us view this first from the perspective of the medical school. For faculty and administrators charged with implementing a medical school's education program, one concern is the logistical quagmire of how to accommodate students moving through the curriculum at varying rates. Consider the potential scenario of every student being at a different point in fulfilling their competencies. Simply tracking each student's progress, coordinating the learning experiences required to develop the knowledge and skills needed to fulfill as yet unmet competencies, and informing preceptors and clinical supervisors of the expected numbers and flow of students would be a monumental task.

But logistics are in some ways the least problematic. Far more difficult is reaching agreement on what should be the trajectory of progression toward proficiency as well as the level of achievement needed to advance to the next level in the educational continuum for the competencies deemed foundational for all medical students regardless of professional career direction. As medical schools wrestle with competency-based progression, it has become clear that a shared mental model of appropriate medical student development does not exist among the faculty within each school. A family medicine or psychiatry faculty member has different expectations for the communication skills required of a medical student at graduation as compared to a radiology faculty member.

In addition, the question at the national level is whether all graduates from all medical schools should be held to the same standard for demonstrating competency achievement. Given different emphases by different medical schools, one can envision that there will be a core set of standard competencies along with additional competencies that are school dependent. While intuitively it makes educational sense to establish uniform graduation requirements for a core set of competencies, defining the core, agreeing upon the criteria for successful completion, and identifying or creating valid and reliable assessment tools is daunting at the school, let alone the national, level.

Initially, I propose that competencies focus on those aspects of clinical practice that will be important whether a graduating medical student goes into neurosurgery, psychiatry, dermatology or any other specialty. While not comprehensive, I have enumerated several in the "Constants section" of this book. While many aspects of future clinical practice are readily identified, it is important to engage a broad constituency of stakeholders in addition to medical educators to identify the foundational competencies. Patients, professionals from nursing and other clinical disciplines, and government stakeholders should contribute to this dialogue. Hopefully, the tendency to define the foundational competencies primarily from a physician centric perspective will be circumvented by broadening the participants.

Coming to agreement on competencies is but the first step. The assessment tools must be identified or developed and the performance standards signifying achievement of competency established. Indeed, one of the impediments to rapid implementation of competency-based progression is the lack of valid and reliable assessment tools. Though innovations in simulation and advances in work-based assessment are promising, much work remains to be done to develop the full spectrum of required assessment tools.

For decades, the task of clinical preceptors has been to assess the developing proficiency of medical students and residents as they take increasing responsibility in the clinical setting. Consciously or unconsciously faculty preceptors judge whether the learner is capable of completing a given task or responsibility independently. A recurring concern is that these professional judgments are "subjective." An interesting development has been the use of entrustable professional activities (EPAs) to assess a learner's trajectory to proficiency. Specific clinical tasks or responsibilities are identified and clinical evaluators are asked whether they would trust the learner to perform the task or take the responsibility in an unsupervised setting. Recognizing that the progression to independence is not all or none, but that there are gradations, entrustability scales enable evaluators to capture their real world judgments regarding a trainee's level of competence [3]. EPAs provide a means to objectify the professional judgment of clinical faculty evaluators.

EPAs have been successfully used across different professions and levels of learners [4–6]. Generally an EPA will map to several competencies for a given individual as EPAs are descriptors of observable tasks or responsibilities. The AAMC has developed a set of 13 entrustable professional activities that tie directly to competencies for graduating medical students entering residency. Examples include "prioritize a differential diagnosis following a clinical encounter," "give or receive a patient handover to transition care responsibility," and "recognize a patient requiring urgent or emergent care and initiate evaluation and management." An accompanying guidebook provides instructions on how to develop and implement EPAs and map to competencies and milestones [7]. Though progress in the development of valid and reliable assessment tools is ongoing, the temptation to default to what can be measured rather than what is important must be resisted.

What cannot be forgotten in the move toward demonstrable achievement is that the most important 'competency' is the maintenance or inculcation of intellectual curiosity and the aptitude for career-long learning, termed "habits of inquiry improvement" by Cooke and colleagues [1]. Not only is this necessary to maintain clinical currency, but as envisioned changes occur, many physicians will need to "reinvent" themselves during the course of their professional careers.

A transition to time-independent, competency-based education at the medical school level is complicated and likely will take many years to accomplish. However, given the positive implications it would have for our medical students and for society, it is imperative that we continue the serious national conversations about how best to move forward.

Much of the impetus for competency-based education came from the Accreditation Council for Graduate Medical Education's (ACGME) initiatives in defining competencies and, through their milestones project, the expected progression of residents through graduate medical education training programs [8]. Although the identification of competencies and expected resident progression as defined by the milestones required thoughtful deliberation and refinement, unlike the situation in medical schools, there is a generally agreed-upon mental model in each specialty of what a competent resident completing their GME program should be able to accomplish. Development of the milestones is a step forward. However, moving to a time independent progression through residency and fellowship programs is unusual and time-based progression remains the norm in the United States.

The Canadian experience is instructive. The University of Toronto orthopedic residency program has successfully implemented a competency-based modular training curriculum, resulting in accelerated resident procedural skill acquisition, diminished downtime, individually-directed times to achieving competency, high resident satisfaction with the program and successful completion of their orthopedics certification examination [9]. The effort required to design, teach and evaluate their residents' knowledge and surgical skills in this new competency-based curriculum was extensive and the logistics were daunting, even with a small number of residents. While recognizing the operational challenges, all Canadian orthopedic surgery training programs and all graduate medical education programs in Canada are transitioning to a competency-based model because of the demonstrable advantages that are expected to positively extend into the trainees' independent professional practices. I anticipate that over time competency-based progression through GME training programs will be adopted broadly across specialties in the United States and in other nations as well. This will be a major disruptor for our academic medical centers and for their clinical services that have come to rely on a consistent number and rotation of residents for care delivery.

Finally, following the lead of the ACGME, enabling policy changes will be necessary from accreditation and credentialing bodies.

Interprofessional and team-based education

A second transformative change in the educational process is interprofessional and team-based education (IPE). The advantages of bringing together the expertise of professionals from multiple disciplines to contribute to the care of the patient have generally been recognized in clinical practice. Acknowledging that interprofessional practice will be a reality for most physicians, IPE has been introduced in many medical schools. However, bringing together students from multiple disciplines for didactic presentations, simulation exercises, and similar curricular events is only a beginning. The most powerful educational tool will be immersing students in clinical environments where they are part of an interprofessional practice that includes learners from other clinical professions.

In most of our academic medical centers, the norm is highly special-
ized clinics and hospital services that do not address the totality of their
patients' problems. While there are notable examples of comprehensive
team-based care in many institutions, such as providing coordinated
care to transplant, pediatric cancer, or amyotrophic lateral sclerosis (ALS) patients and their families, the numbers of medical student
learners who can be immersed in these services is limited. Moreover,
the focus is often on complex and devastating disease processes rather
than comprehensive approaches to the care of patients with common
chronic diseases. For example, the patient with congestive heart failure not infrequently also has diabetes mellitus, mobility limitations due
to osteoarthritis, may have cognitive decline, potentially has difficulty
with transportation and getting to clinic visits, may have financial limitations that impact his ability to purchase medications, and lives in an
area where personal safety concerns limits his options for exercise and
leads to social isolation. The complexity of issues facing such patients
is beyond the skill set of any physician and a multi-professional team
approach is needed to optimize the individual's care and their ability
to adhere to complicated medical regimens. Even when high performing teams comprised of professionals from multiple disciplines are involved in the patient's care, they are often geographically distributed
and asynchronous in their patient interactions so their contributions are
not readily apparent nor an integral part of the medical student's educational experiences. As importantly, whether it be the medical students
or students from other professional disciplines, the learners are usually
attached to the clinical team member from their discipline and do not
have the benefit of in-depth learning with faculty and students from
other professions.

Consequently, many medical students and residents are only peripherally aware of the expertise that other professionals such as social workers,
physical therapists, occupational therapists, speech therapists, pharma-
cists, and psychologists possess and contribute to a patient's care. Medical
educators need to ensure that students understand the skill sets of other
healthcare professionals and how the appropriate team of professionals
enhances patient care and outcomes. This requires that students are immersed in clinical contexts that include multiple clinical disciplines practicing collaborative team based care consistent with practices best suited
to optimize health and minimize disease morbidity.

A less frequently acknowledged but very important lesson will be the realization that teams of professionals, even those united around the laudable
goal of enhancing the outcomes for patients, will look at the same patient
situation and frame the problem and potential solutions differently. There
will be disagreements regarding the best course forward. There will be different agendas and priorities among the team members. Simply assembling

a group of people from different healthcare disciplines is not a panacea. The hard work of teamship is ongoing and requires constant attention.

As expressed by Lingard

Use interprofessional practice/interprofessional education to focus on relationships among the parts, not the parts themselves.
Build complexity in: role overlap, negotiated authority, competing motivations.
Move training into practice settings.
Develop faculty ability to constructively reflect on practice tensions with learners [10].

This requires a willingness, not only on the part of medical educators but also the cooperation and commitment of administrators and professionals from the involved clinical disciplines to change long-standing care and educational patterns in an effort to create better patient care and learning environments. Creation of such clinics and services is necessary if the promise of interprofessional practice leading to better patient outcomes is to be realized. Limiting IPE to classroom and simulation venues will send a strong message to students that interprofessional practice is not part of the "real world" of clinical medicine.

Educational practices deserving strengthening

Despite all the advances in clinical medicine, our increased understanding of the mechanisms of health and disease, and the potential of technology to facilitate education, the most important teacher is the patient. While standardized patients and sophisticated technologic simulations are useful educational tools, they cannot replace meaningful patient engagement as an essential educational experience.

As expressed by Sir William Osler over 100 years ago,

To study the phenomena of disease without books is to sail an uncharted sea, while to study books without patients is not to go to sea at all. [11]

Through patient care and interaction, the learner's seemingly disparate knowledge gained through coursework and prior experiences is integrated into an understanding of disease presentation, the natural history of disease progression, and the approach to diagnosis and therapeutic options. Their understanding of the effect of illness and disability on individuals, their families and social network is deepened. The complexities of the clinical care system are experienced, and skills are developed for navigating the many facets of the system on behalf of one's patients.

Almost a century ago, at a time when clinical learning was essentially limited to the hospital, Francis Peabody in a lecture to Harvard medical students expressed,

> ... the study of medicine in the hospital actually becomes the practice of medicine, and the treatment of disease immediately takes its proper place in the larger problem of the care of the patient. [12]

Although the clinical context for learning has expanded to include ambulatory and community sites, the centrality of patient care for the education and professional development of students and trainees remains a truism.

Let us consider some additional educational practices and opportunities that should be used to enhance our medical students' and residents' skills and knowledge.

Clinical teaching

At one time, work rounds on every inpatient clinical teaching service in academic medical centers included the entire clinical team—attending, residents, and students. As each patient on the service was seen, an update on the patient's progress would be presented by a resident or a student, a pertinent interval history and physical examination would be performed by the attending calling attention to important findings, and a discussion as to next steps and key teaching points would ensue in the presence of the patient. Currently, work rounds are oftentimes conducted in a conference or staff room rather than at the patient's bedside. While often times more convenient and efficient, it eliminates the opportunity for the medical students and residents to observe their senior faculty members role model interactions and discuss clinical issues with patients [13].

Even more concerning is the trend of inpatient rounds to become focused on the information available in the electronic health record rather than the patient. This evolves into a discussion of the efficient movement of patients through a proscribed clinical pathway rather than a thoughtful assessment of the patient's situation that provides an opportunity for learning.

Conference room discussions, chart rounds, computer rounds, all in the name of efficiency, along with the tyranny of length of stay, have all too often replaced bedside work rounds that included key teaching points and discussion. The inpatient setting remains a valuable venue for teaching, but medical educators need to consider how best to maximize its learning potential and the lessons being conveyed, especially for novice and intermediate learners.

"…I don't think efficiency is the number one thing for a really good doctor who is thoughtful. You can become more efficient, but we train them (students and residents) and we reward efficiency. Like, a really good student is one that can do everything and is super-efficient. But the problem with that is you get somebody that's very efficient but it might not be good efficiency. It might be efficiency at mediocrity and that's not really what most people (patients) want. You don't just want stuff done for you, you want the right things done and you want the doctoring and the healing and all the other things that go in the doctor-patient relationship and we are losing that in our worship of efficiency."

Doug Paauw

The erosion of clinical teaching is not limited to the hospital setting. Though patients are increasingly cared for entirely in the ambulatory setting, the constraints inherent in an ambulatory clinic require creative approaches to education. Well known examples include the One-Minute Preceptor model that employs five "microskills" [14, 15].

1. Get a commitment—the learner states his/her diagnosis or plan
2. Probe for supporting evidence—assess the learner's knowledge supporting their plan and their reasoning
3. Teach general rules—provide specific "take-home points" that are applicable to future cases
4. Reinforce what was done well—giving positive feedback
5. Provide feedback for improvement with specific recommendations.

Notwithstanding the creation of teaching methods tailored to the realities of ambulatory care, the growing expectations for numbers of patients seen by a faculty member plus the burden of documentation to meet arbitrary quality metrics and billing requirements, means that even the most motivated clinicians often need to limit or eliminate their medical student and even their resident teaching responsibilities. Ensuring that our motivated and excellent ambulatory teaching faculty remain engaged in the medical education process will be vital as ever more change occurs.

Faculty expertise and specialty services

Beyond the stresses placed on the clinical learning environment by efficiency and performance demands, the increasingly narrow and sub-specialized expertise of faculty, coupled with the proliferation of inpatient specialty units and specialty ambulatory clinics, has compromised the educational value of many clinical settings. Novice learners, which is what medical students are, develop their mental models of the presentation and course of different illnesses through their interaction with patients. When discussions of patients are focused strictly on the problem at hand, rather than considering the spectrum of potential concerns, incomplete mental models may develop.

For example, the differential diagnosis of a middle-aged woman with ascites on an inpatient internal medical service will include alcoholic liver disease. Ovarian cancer is less likely to be discussed and may not even be considered. Conversely, on the inpatient gynecologic oncology service, the discussion will focus on ovarian cancer and the possibility of alcoholic liver disease will rarely arise. Moreover, the attendings on the respective services are often uncomfortable teaching about diseases and illness presentations outside their area of expertise. While providing sophisticated clinical care for the most complex patients, the sub-specialization of physicians and services in most academic medical centers is not optimal for the education of novice medical students. A growing dilemma in our modern academic medical centers is "who will teach?" [16].

Just as importantly, clinical venues are needed where medical students and residents can care for and learn from a variety of patients at different stages in their disease presentation and progression. This experience will allow the learners to develop the robust mental models that will be foundational for their professional careers.

Grand rounds

Though much of the development of clinical reasoning and procedural skills occurs in small group or even 1:1 interactions, additional venues provide the opportunity to deepen understanding and enhance learning. While recognizing that the march of progress may obviate the need for some educational exercises, select traditional venues, with thoughtful modifications, remain important options.

Let me share a personal experience. As a guest at Peking Union Medical College in Beijing, I was privileged to attend Internal Medicine Grand Rounds. The auditorium was filled to capacity as it was the expectation that all faculty, residents and students would attend. The Chair of the Department of Internal Medicine was seated in the front row, center seat, flanked by his Division Chiefs. The patient to be discussed was brought into the auditorium and his history, laboratory and imaging studies were presented. Subsequently the audience was invited to come on stage to examine and question the patient.

Briefly, this was a man in his late 30s who had progressive weakness and lethargy over several months, a history of smoking 2+ packs per day, marked hypercalcemia, renal failure, a history of tuberculosis with multiple lesions on chest X-ray and generalized osteopenia. Without antecedent preparation or warning, the Chief Resident then asked the pulmonary, renal, endocrine and metabolic, and infectious disease Division

Chiefs to come and reason through the case, including steps they would take to make a diagnosis, while explaining their thinking to the audience. Subsequently, questions and alternative diagnostic possibilities were entertained from the audience.

As the patients that were presented at Grand Rounds had not been diagnosed at the time of presentation, the chief resident then gave updates on cases that had been presented in preceding weeks. While I am certain that I missed some of the nuances of the discussion in translation, I was most impressed that students, residents, and faculty were afforded the opportunity to understand how the most senior and experienced clinicians approached clinical problems.

Peking Union Medical College was funded by the Rockefeller Foundation in 1917 and modeled after Johns Hopkins Medical School. Prior to World War II, there had been a continuing presence of young Hopkins faculty, many of whom would go on to distinguished careers and leadership positions at Johns Hopkins. The Grand Rounds I was privileged to witness was how grand rounds were conducted in US academic medical centers through the mid-20th century. While not minimizing the importance of understanding the scientific basis of disease, when considering optimal ways to teach the practice of clinical medicine such a learning exercise is superior to the erudite scientific discourse that oftentimes is the hallmark of grand rounds. Whereas presentations on the latest scientific understanding and clinical applications thereof to specific clinical problems and diagnoses is of value to faculty and senior residents, it is less so to novice and intermediate level learners. Medical educators need to thoughtfully match the educational content to the intended audience.

Professional identity

We know that direct interaction with patients will help students develop their disease oriented clinical reasoning abilities. However, as important to learning the application of the science of medicine to clinical care is developing an appreciation for and expertise in the art of medicine. Fundamentally, clinical medicine is a human to human interaction. In the clinical setting, students and residents have the opportunity to develop and refine their communication skills, develop an appreciation for their patients as unique individuals, and interact with patients and families from widely differing racial, ethnic, religious and social economic backgrounds. This is where attitudes—respect, empathy, dedication to one's patients—are learned and reinforced. The power of the "hidden curriculum" to strengthen or extinguish positive behaviors and attitudes in the nascent physician is incontrovertible.

Again, this necessitates attention being paid to the learner's clinical team (including non-physician professionals) and the spectrum of patients they see. Do the faculty, nurses and other team members exhibit the professional behaviors and interactions one with another as well as with patients and their families that we want our learners to develop? Do they take the time to know their patients as individuals or merely see them as "the elderly man with congestive heart failure in room 237?" Do they actively involve patients and family in care decisions as appropriate? The importance of positive role models who have the scientific and clinical knowledge to provide the highest quality of care and consistently exhibit the highest professional standards is unquestioned.

Indeed, the totality of what it means to be a physician, their professional identity, is developed through students' and residents' immersion in the clinical environment. In their 1991 book *Situated Learning: Legitimate Peripheral Participation* [17] Lave and Wenger set forth that professional learning is not simply the accumulation of a series of facts and skills. Rather, it is a process of socialization or enculturation. The novice performs simple but important tasks as part of the "team." As expertise is developed, greater responsibility is given and they become ever more central to accomplishing the tasks of the discipline or profession, until they become full-fledged participants.

Medicine was a prime example of this immersive situated learning. For decades medical students were assigned important but simple tasks such as tracking down x-rays, drawing bloods, inserting IV lines and catheters—SCUT. As they progressed into their fourth year, they were usually required to complete a sub-internship where, similar to a postgraduate year 1 (PGY1) resident, they assumed the role of the primary "physician" responsible for the care of patients while under the watchful supervision of faculty. Progressing into their PGY1 year, they cared for increasing numbers of patients and, over time, as senior residents, they supervised junior residents and were charged with the responsibility for caring for a service. Through this progression of increasing responsibility, they developed their clinical acumen, their attitudes toward patients and other care professionals and the leadership skills necessary to lead and to be a full-fledged member of the hospital-based clinical care team. They became socialized into what it meant to be a physician.

This progressive increase in responsibilities has been disrupted. As noted previously, the unintended consequences of regulatory changes, the widespread implementation of ancillary clinical support services, advances in electronic records and transmittal of clinical information, and the emphasis on decreasing length of stay has limited the participation and contributions of medical students and even residents, potentially restricting their learning opportunities and slowing professional development.

I am not arguing that there should be a return to the days of medical students functioning as "scut monkeys." Rather that medical educators must be attuned to professional identity formation and recognize that students' complaints of not being involved in meaningful activities and feeling like a "vestigial appendage" are symptomatic of a potentially deeper problem. It is imperative that faculty thoughtfully involve medical students in significant functions of the clinical care team and that residents are ceded ever greater responsibilities based on their professional maturity and skill sets. This requires creativity. Attention to professional identity formation is as important as ensuring the adequacy of students' and trainees' knowledge and skill bases.

Deliberate practice and feedback

In order to develop expertise, deliberate practice, and not simply time on task, is required. This necessitates that faculty members develop skills in providing meaningful feedback that can be utilized by the learner to improve performance [18, 19]. While the important role of formative feedback is widely recognized, a recurring theme from students is that the feedback they generally receive is unhelpful. Platitudes such as "read more" or "good job" are not useful. The specificity needed for "deliberate practice" is absent.

To address this issue, several things must occur. Faculty must observe the student or resident sufficiently to be able to specifically reinforce positive behaviors and provide guidance on areas needing attention. This requires changes for the already overcommitted faculty both in mindset and in practice. Faculty must develop trust in the validity of their assessments. Unfortunately, rather than recognizing an experienced faculty member's assessment of learners as expert judgment, it is usually referred to as "subjective" which implies that it is of little value as compared to "objective" assessment. I would argue that the experienced teacher develops the ability to assess learners just as the experienced clinician develops the ability to make diagnoses. Rather than being dismissive of this "subjective" assessment, the focus should be on translating for the learner what needs to be done to improve. A recurring argument is that such assessments are prone to bias. While this may be true in some cases, the fact is that "objective" assessments such as multiple-choice tests are also biased in their sampling.

In addition, faculty need to have confidence to provide negative as well as positive feedback and how to provide input that will be considered as valid by the learner.

In turn, medical students and residents must be immersed in a learning environment where it is safe to acknowledge deficiencies in order to strengthen one's performance. The all too common effort to maintain a veneer of proficiency in order to impress the attending or the resident should be antithetical to the learning environment. Students and residents will need to change and embrace the vulnerability needed to admit their deficiencies.

Moreover, the inherent distrust of "subjective" evaluations must change. The majority of one's professional career is spent in settings where feedback is nonspecific. Although patient ratings, 360-degree feedback, and quality metrics are increasingly common, even with these "objective" feedback mechanisms the physician is left to interpret how best to improve upon their performance. Much more common, and likely to continue to be so, is the nonspecific and oftentimes indirect feedback received through interactions with patients and colleagues. Being given specific actionable feedback is rare especially once one is in professional practice. Students must learn that feedback often comes in the form of subtle nonspecific cues. This is part of their journey to physicianhood.

Medical educators will need to be seen as partners with their learners on the journey to full-fledged independent professionals. This requires vulnerability on the part of faculty and learners alike. It requires frame shifts in how each perceives their roles. And, it will require a level of engagement that is not readily achieved in our frenetic clinical environments.

The journey forward

Medical student education must focus on establishing the foundation for a lifelong professional practice. As noted, our goal is to develop physicians who are autonomous self-directed learners. Lifelong learning has been recognized as essential for physicians for decades. However, the reality to date has not been encouraging. Potentially, this requires medical educators to change their approach. Rather than focusing primarily on learning content and skills, learning how to learn, termed heutagogy [20–22], should have equal emphasis. I have provided a few examples of refocused, expanded and completely new domains that will be important for students to understand if the directional changes in the practice of clinical medicine continue apace. I fully anticipate that there will be additional new knowledge domains and skills that will become important due to unanticipated technologic advancements, scientific discovery, and unforeseen disruptors, hence the primacy of learning how to learn.

From pedagogy to andragogy, and onwards to heutagogy

As presented in a 2015 OECD report "…many educational establishments consider learning in terms of content delivery… They set a curriculum and work to it, ensuring that no content is left out …. This is the traditional domain of pedagogy – predetermining what the learning outcomes will be and filling all the perceived gaps on behalf of the learner.

… andragogy, where there is a degree of self-determination on behalf of the student. … self-direction increases motivation and can be used to demonstrate self-reliance and the ability to spot opportunities for students' own learning … (i) move from dependency to self-directedness; (ii) draw upon their reservoir of experience for learning; (iii) are ready to learn when they assume new roles; and (iv) want to solve problems and apply new knowledge immediately.

The <u>heutagogical</u> stance takes things further … to develop self-reliant students who see opportunities in problems for themselves, and have the means to develop their own approaches … within this conceptual progress from pedagogy to heutagogy, "knowledge giving" by an expert becomes outdated … one of the main aims of the approach is to help to develop learner confidence and self-efficacy – to better enable them to do things themselves, as well as to simply understand. …

…pedagogy is teacher-centered and teacher oriented, andragogy is more learner-centered. Finally, heutagogy is student-determined objective setting that is developed as a result of being opportunity aware" [23].

Implicit in this journey from pedagogy to heutagogy is a fundamental redefinition of what it means to be a "teacher." While the passage of the "sage on the stage" is rued by some faculty, the diminishing role of large lectures in medical education is a positive trend, assuming that the focus is on learning, not just for the present, but truly "learning to learn." The primacy of the content expert needs to be augmented or even supplanted by teams of faculty who combine subject matter knowledge with technologic and data analytic expertise. Increasingly, the role of the faculty member must be to provide guidance, feedback, and coaching. Unfortunately few medical school faculty members have the well-developed skills to fulfill these roles. The magnitude of the faculty development efforts required to make the transition to new roles and to fully capitalize on emerging learning technologies models must not be underestimated. This change will be disruptive but necessary. The long-term results for our students and residents will be worth it.

To develop the clinical experiences necessary for immersive interprofessional education, to mount the faculty development efforts needed to equip faculty for their new educational roles, to carve out and protect sufficient time for faculty to teach and mentor, and to reassert that the education of learners at all levels is central to the mission of an academic medical center will require engagement of not only medical educators but also the willingness and involvement of academic leaders and administrators.

References

[1] Cooke M, Irby DM, O'Brien BC. Educating physicians: a call for reform of medical school and residency. San Francisco, CA: Josey-Bass/Carnegie Foundation for the Advancement of Teaching; 2010.
[2] Lucey CR. Achieving competency-based, time variable health professions education. In: Proceedings of a conference sponsored by Josiah Macy Jr. Foundation in June 2017. New York, NY: Josiah Macy Jr. Foundation; 2018.

[3] Rekman J, Gofton W, Dudek N, Gofton T, Hamstra SJ. Entrustability scales: outlining their usefulness for competency-based clinical assessment. Acad Med 2016;91(2):186–90.

[4] ten Cate O. Entrustability of professional activities and competency-based training. Med Educ 2005;39(12):1176–7.

[5] Elnicki DM, Aiyer MK, Cannarozzi ML, Carbo A, Chelminski PR, Chheda SG, et al. An entrustable professional activity (EPA)-based framework to prepare fourth-year medical students for internal medicine careers. J Gen Intern Med 2017;32(11):1255–60.

[6] Mulder H, ten Cate O, Daalder R, Berkvens J. Building a competency-based workplace curriculum around entrustable professional activities: the case of physician assistant training. Med Teach 2010;32(10):e453–9.

[7] Association of American Medical Colleges. Drafting panel for core entrustable professional activities for entering residency. Core entrustable professional activities for entering residency. Curriculum developers' guide. Association of American Medical Colleges; 2014.

[8] Edgar L, Roberts S, Yaghmour NA, Hunderfund AL, Hamstra SJ, Confroti L, et al. Competency crosswalk: a multispecialty review of the Accreditation Council for Graduate Medical Education milestones across four competency domains. Acad Med 2018;93(7):1035–41.

[9] Ferguson PC, Kraemer W, Nousiainen M, Safir O, Sonnadara R, Alman B, et al. Three-year experience with an innovative, modular competency-based curriculum for orthopaedic training. J Bone Joint Surg 2013;95(21):e166. https://doi.org/10.2106/JBJS.M.00314. Series A.

[10] Lingard L. Are we training for collective incompetence? Common education assumptions and their unintended impacts on healthcare teamwork. April 2 2019. Ann Arbor, MI.

[11] William Osler Quotes. BrainyQuote.com, Xplore Inc. https://www.brainyquote.com/quotes/william_osler_394466; 2018. (Accessed February 22, 2018).

[12] Peabody FW. The care of the patient. JAMA 1927;88(12):877–82.

[13] Cruess SR, Cruess RI, Steinert Y. Teaching rounds: role modeling: making the most of a powerful teaching strategy. BMJ 2008;336:718–21.

[14] Neher JO, Gordon KC, Meyer B, Stevens N. A five-step micro skills model of clinical teaching. J Am Board Fam Pract 1992;5:419–24.

[15] Furney SL, Orsini AN, Orsetti KE, Stern DT, Gruppen LD, Irby DM. Teaching the one minute preceptor. J Gen Intern Med 2001;16:620–4.

[16] Woolliscroft JO. Who will teach? A fundamental challenge to medical education. Acad Med 1995;70(1):27–9.

[17] Lave J, Wenger E. Situated learning: legitimate peripheral participation. Cambridge: Cambridge University Press; 1991.

[18] Bing-You R, Hayes V, Varaklis K, Trowbridge R, Kemp H, McKelvy D. Feedback for learners in medical education: what is known? A scoping review. Acad Med 2017;92(9):1346–54.

[19] Tekian A, Watling CJ, Roberts TE, Steinert Y, Norcini J. Qualitative and quantitative feedback in the context of competency-based education. Med Teach 2017;39(12):1245–9. https://doi.org/10.1080/0142159X.2017.1372564.

[20] Canning N. Playing with heutagogy: exploring strategies to empower mature learners in higher education. J Further High Educ 2010;34(1):59–71. https://doi.org/10.1080/03098770903477102.

[21] Blaschke LM. Heutagogy and lifelong learning: a review of heutagogical practice and self-determined learning. Int Rev Res Open Dist Learn 2012;13(1):56–71.

[22] Reem RA, Ramnarayan K. Heutagogic approach to developing capable learners. Med Teach 2017;39(3):295–9. https://doi.org/10.1080/0142159X.2017.1270433.

[23] Penaluna A, Penaluna K. Entrepreneurial education and practice. Part 2—building motivations and competencies. OECD; 2015. p. 14–5.

Additional resources

- Taylor DCM, Hamdy H. Adult learning theories: implications for learning and teaching in medical education: AMEE guide no. 83. Med Teach 2013;35(11):e1561–72. https://doi.org/10.3109/0142159X.2013.828153.
- Lucey CR. Achieving competency-based, time variable health professions education. In: Proceedings of a conference sponsored by Josiah Macy Jr. Foundation in June 2017. New York, NY: Josiah Macy Jr. Foundation; 2018.
- Mitchell P, Wynia M, Golden R, McNellis B, Okun S, Webb CE, et al. Core principles & values of effective team-based health care. [Discussion paper]. Washington, DC: Institute of Medicine; 2012. https://nam.edu/perspectives-2012-core-principles-values-of-effective-team-based-health-care/.
- National Academies of Sciences, Engineering, and Medicine. Strengthening the connection between health professions education and practice: proceedings of a joint workshop. Washington, DC: The National Academies Press; 2019. https://doi.org/10.17226/25407.
- National Academies of Sciences, Engineering, and Medicine. A design thinking, systems approach to well-being within education and practice: proceedings of a workshop. Washington, DC: The National Academies Press; 2019. https://doi.org/10.17226/25151.

CHAPTER

19

Considerations when designing educational experiences

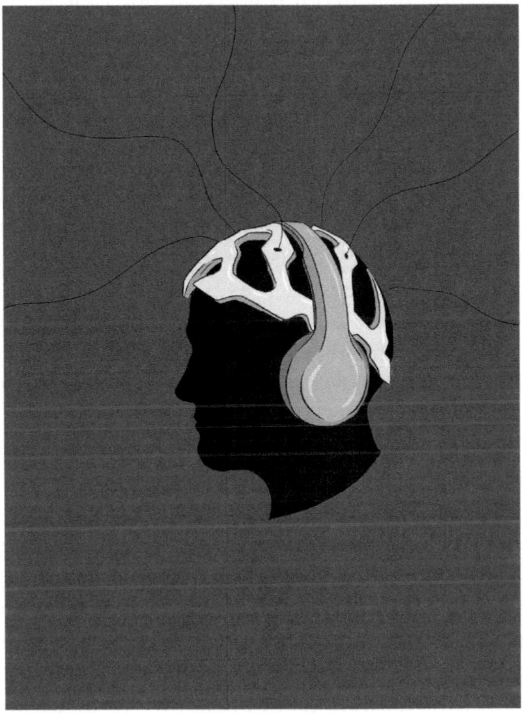

I never teach my pupils, I only attempt to provide the conditions in which they can learn. ***Albert Einstein***

While medical educators do not have direct control over the external governmental actions and policies that have played a major role in shaping our academic medical centers and continue to impact the continuum of medical education, they do control how they respond and adapt to the

Implementing Biomedical Innovations into Health, Education, and Practice
https://doi.org/10.1016/B978-0-12-819620-5.00019-9
© 2020_ James O. Woolliscroft. Published by Elsevier Inc.
All rights reserved.

ever-changing educational environment. As advances occur in scientific understanding and the practice of clinical medicine, priorities regarding content, instructional methodologies and even the clinical context where learning takes place have changed. While often seemingly incremental, when viewed over time, the content, process and the tools employed to enhance learning across the education continuum have evolved. This is responsible medical education. It is expected to continue, and at an accelerated pace.

Rather than being purely reactionary, as history has shown us, visionary medical educators can also shape the future. This requires not only an understanding of the core tenets of our profession but also an appreciation and knowledge of advances in disciplines outside of medicine that are relevant to the educational process. In this chapter we will consider a few examples of developments in our understanding of learning that have potential application to the training of future physicians.

Testing effect

One example is the body of published work that examines ways to strengthen long-term memory. Stating the obvious, generations of students have proven that studying is the foundation for learning medicine. However, common techniques such as highlighting, rereading, and summarizing notes are variably efficacious. Augmenting studying by requiring learners to retrieve information through frequent testing—the "testing effect" or "retrieval practice effect"—has been shown to be more effective than simply reviewing material [1]. Intuitively, students recognize the value of practice questions and "quizzing" themselves as useful means to enhance their learning. Not surprisingly, "test questions" feature prominently in commercial board review courses.

However, the importance of empiric studies of the "testing effect" as applied to medical education is evidenced by studies suggesting that its value decreases as the complexity and numbers of interacting information elements increase [2]. As much of the work establishing and confirming the "testing effect" has been done under laboratory conditions or with discrete domains of information, its applicability to medical education that deals with complex and interrelated content needs to be established.

Learning order

The order in which information or skills are best learned has been the subject of intense study. When considering three types of learning elements (A, B, and C), is it best to learn each one to a level of proficiency before moving on to the next (AAA, then BBB, then CCC—known as blocked) or mixing the three and learning them together (ABCABCABC—known as interleaved)? Studies show that while a blocked approach is more efficient for short-term results, interleaving enhances "long-term" performance in widely varying contexts, ranging from college baseball players practicing hitting different types of pitches, to identifying paintings by artist, to mathematics [3]. Might there be implications for how material is presented in many of our medical schools? While discipline-based courses have evolved into organ system curricula, the specific organ system sequences are frequently standalone with little purposeful linkages across organ systems. Might there be long-term learning enhancement by intentional admixture or interleaving of knowledge and skills? Indeed, this is the concept underlying the integrated, spiral curriculum [4].

Forgetting curve and spacing

An intriguing question is whether the observed positive effects of interleaved practice are actually a reflection of spacing. Building on work done by Ebbinghaus describing the forgetting curve in 1885 (Fig. 19.1), we know that learners will forget on average 90% of material within 1 month [5]. We also know that revisiting material at regular intervals, whether through presentations, electronic communication, or testing, enhances retention (Fig. 19.1). In a randomized controlled study incorporating spacing principles, a validated web-based teaching program on four core urologic topics for third-year medical students on a one-week clinical rotation was followed by a series of 11–13 weekly educational emails consisting of a short,

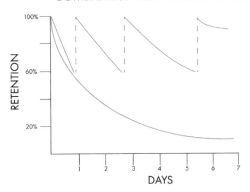

FIG. 19.1 Effect of time on retention of learning—the forgetting curve—and the salutary effect of quizzing and structured interventions on retention.

clinically relevant question followed by the correct answer and teaching points. While all four topics were covered in the follow-up educational emails, students were randomized to receive weekly emails in only two of the four urologic topic areas. An end-of-the-year test of students who had completed their urology education 6–11 months earlier showed significant improvements in the retention of clinical knowledge in the two topic areas covered by the emails they received [6]. In a similar study of histopathology diagnostic skills of hundreds of US urology residents, online spaced education substantially improved their long-term retention [7]. While these studies were in the context of a single clinical discipline, similar techniques could be applied broadly. If verified in additional clinical contexts, study is required to determine optimal spacing and assessment as well as the best techniques to revisit material. Just as these studies explore spacing principles that have been empirically observed to enhance learning in laboratory and other educational contexts, it is important to rigorously study all educational methods and principles that have been derived in other domains to determine their relevance in medicine [8].

Spaced learning, retrieval practice and interleaved learning are but a few of the multiple learning principles and theories potentially applicable to medical education [9]. As is increasingly recognized, it behooves medical educators to intentionally integrate education theories in our educational research and practice [10]. Even as the relevance of innovations based on sound educational theory is proven in medicine, the challenges of implementation are not to be underestimated, as our courses, clerkships and rotations often exist in isolation one from another. Because the body of knowledge and the number of required skill sets is immense, making full implementation across medical education very complicated, it will be important to start by applying these learning best practices to discrete units before attempting to scale up.

Threshold concepts

Experienced medical faculty often recognize that there are certain concepts within a discipline that students have difficulty grasping. Moreover, until these concepts, termed threshold concepts, are mastered further understanding is impeded, and anxiety and confusion often results. Characteristics of threshold concepts include:

- transformative: they shift the learner's perceptions of a subject
- irreversible: once learned, they are hard to unlearn
- integrative: they expose the interrelatedness of things, the "aha" moment

- bounded: coupled with other threshold concepts they define a disciplinary area
- troublesome: they are difficult and not intuitive for learners.

While generally agreed upon, how many of these characteristics are required to define a concept as threshold remains debatable. Simplistically speaking, threshold concepts are core understandings necessary for a student to progress. But they are distinct from core concepts that are necessary to understand a subject but do not transform one's thinking about a subject.

Within medicine, threshold concepts are as basic as learning how to interpret a three dimensional structure from the two dimensional representation on an image [11].

For medical students on their initial clinical experiences this means accepting:

- responsibility for active self-directed learning,
- responsibility for their patient's care,
- that physicians have multiple roles,
- that uncertainty is inherent in medicine,
- that medicine requires a strategic rather than formulaic approach,
- understanding the context and complexity of the patient is necessary to manage a medical problem,
- that effective care requires patient centeredness and a team of clinical professionals,
- that documentation is essential [12].

The fragmentation of our curriculum and learning experiences exacerbates students' difficulties making connections among disciplines and fundamental concepts [13]. Furthermore, as noted in the definition, threshold concepts are irreversible and transformative. Expert faculty who have traversed this threshold may find it difficult to understand their students' struggles, let alone be able (or willing) to guide them through the process. Unfortunately students may have a superficial understanding and be able to "pass" an assessment but then be stymied in their progression.

Potentially, learning analytics will be valuable tools to identify threshold concepts with which students struggle. In the meantime, medical educators have to rely on observation of their learners to identify areas of uncertainty and anxiety potentially indicative of a threshold concept. Sharing insights among faculty, gained from experience working with learners at different levels, will assist in prioritizing areas needing special attention to ensure that all students grasp these transformative concepts.

IV. Implications for educators

Learning analytics

Learning analytics, applying statistical tools to the ever increasing amount of data available on our learners, has the potential to inform trainee assessment at an individual level and identify educational program problems. The wide array of data that is being collected on our learners, ranging from admissions information, to exam performance, to daily encounter cards, to progress testing, to milestones, to entrustable professional activity ratings, and more, provides the opportunity to examine not only outcomes but also the educational process.

Importantly, while analyses of such large data sets may provide important insights, they can also produce "patterns" that are meaningless. Applicable to learning analytics are the cautionary reminders regarding machine learning generated algorithms and clinical medicine. Bias, missing data, measurement error and the statistical method or program used are but some of the considerations when interpreting the findings. As the data sets grow, "statistically significant findings" will be present. The key will be identification of the meaningful ones. Are the relationships plausible? Do they fit with educational theory? Is prior experience with learners consistent with the findings?

Undoubtedly learning analytics will prove to be a powerful tool to enhance our assessment of individual learner's progress as well as our educational programs. However, the path to this understanding will not be linear and educators must be informed consumers of the product.

Cognitive load theory

Cognitive load theory, first introduced in the 1980s as useful for informing instructional design, attempts to explain how information processing load affects students' abilities to learn new information and construct long-term memories. The central thesis is that limitations of working memory constrain the amount of novel information that can be processed. In contrast, there are no known limits when familiar organized information is processed from long-term memory. Cognitive load theory focuses on the demands on working memory as students are learning novel information. The complexity of the material, how information is presented including the pace of information presentation and extraneous environmental distracters all reduce the working memory resources required to learn.

Refinements of the theory and implications for instructional design have been ongoing for decades. One such model is the four component instructional design (4C/ID) [14]. Useful for the development of professional competencies, it assumes that complex skills include "recurrent skills" used over multiple tasks in addition to "nonrecurrent" skills which require problem-solving and reasoning. Applied to real life tasks, learners initially work on simple problems and as expertise is acquired graduate to more complex problems while simultaneously receiving less guidance and support. The spiral curriculum and legitimate peripheral participation are examples.

As research evidence has accumulated regarding the importance of cognitive load for successful and efficient learning, when developing educational programs consideration of the cognitive load imposed by the instructional design seems reasonable. However, how does one measure cognitive load? Subjective measures ask learners to rate the perceived difficulty of a task. Objective measures include eye tracking and pupillometric analysis, and physiologic data such as heart rate. Secondary task analyses measure performance on two tasks such as learning and rhythmic foot tapping as both require cognitive effort [15]. However, the applicability of any of these measures, other than the subjective ratings by students, to educational offerings is limited to research investigations. That said, the concept of cognitive load is worthy of consideration when developing new educational experiences.

Adult learning theories

The entire educational spectrum, from premedical through continuing professional development, is all about learning. But how do adults learn? Are there ways to enhance learning? To make learning more efficient?

There are multiple theories in several broad categories that attempt to explain how adults learn:

- *Instrumental learning theories* including the behaviorist and cognitive learning theories that focus on individual experience.
- *Humanistic theories* focusing on self-actualization and individual development.
- *Transformative learning theory* in which critical reflection is used to challenge beliefs and assumptions.
- *Social learning theories* that emphasize communities of practice shaping the learners development.
- *Motivational models* that recognize intrinsic motivation sustained by autonomy, competence, and a sense of belonging or relatedness.
- *Reflective models* that propose that reflection leads to action and change [9].

Is there a practical relevance to "real world" medical education of these theories? Rather than focusing on a specific theory, combining the salient features will serve as useful guides for medical educators.

Although medical students matriculate with an extensive array of experiences and backgrounds, a key to success is providing them a way to organize the information they will be learning. Hence, the importance of a mental model or scaffold to serve as a guide as they attempt to organize and comprehend the volume of complex material they will encounter. Students will need to overtly connect new information to previously learned concepts and knowledge and then be encouraged to consider additional potential connections and relationships. This reflection can be facilitated through discussions, demonstrations, seminars and even informal conversations. Then the learner should be helped to reflect on the strengths and weaknesses of their understanding through assessments that can be formal or informal.

These steps are obvious but may not always be followed in working with medical students and residents. In addition, learning theories provide perspectives on motivation, the context in which learning occurs, and a host of other factors. It is important, however, to remember that scientists involved in education research are interested in developing theories that may or may not translate directly into practical applications. Medical educators should be informed of progress in the development of learning theories but need to consider the context, sophistication of the learners they are working with, and the desired outcomes when considering the applicability of a theory to their educational situation.

Educational effectiveness studies

Just as there has been an increasing focus on the translation of controlled clinical trials to clinical practice, clinical effectiveness studies, so too there needs to be testing of educational theories, educational effectiveness studies. Laboratory research and tightly controlled experiments have led to many of the recommendations regarding enhancement of learning. Before assuming that findings of efficacy under controlled experimental conditions translate directly into enhanced effectiveness of our medical education programs, further study is necessary.

Many randomized controlled clinical trials have shown positive efficacy outcomes under optimal circumstances but were not as effective when examined under real-life conditions. Similar findings have been noted in studies of educational interventions. In one example, a trial used an intervention that had been shown to improve communication skills as assessed by standardized patients. However when nurses and residents were studied in a randomized controlled trial comparing the simulation-based skills training with "usual education," patient and family assessments of the quality of communication showed no difference between the two groups. Surprisingly, patients reported increased depressive symptoms in the intervention group [16]. This highlights the importance of subjecting even those interventions that are thought to be educationally sound to appropriate study.

I recognize that there are multiple difficulties with such a proposal. Educational placebos are nonexistent. The educational milieu is complex with many factors that cannot be controlled. Often times learners feel disadvantaged if they do not receive an intervention equivalent to another group. Desired outcomes from the educational intervention may take years to materialize. These are but some of the complexities inherent in medical education studies seeking to establish how educational theories and research developed under optimal conditions perform in "real life" educational situations [8].

Technology tools

Even though challenges remain in the implementation of spaced learning, frequent testing and similar concepts, the tools available for medical educators to facilitate such innovations have never been greater. Already, medical students and residents are adept at finding e-learning programs, videos, and other resources that fit their learning needs. Oftentimes faculty are unaware of all the available high-quality e-learning resources that could be incorporated into the curricula. Even when faculty members know what the learners actually use, faculty recommendations frequently do not align with student rankings of what they find most valuable. While our learners are adept at accessing and utilizing electronic learning aids, faculty still tend to think of textbooks (the irony!) and published articles as superior sources of information.

The potential of technology to revolutionize instruction is both appealing and disturbing. Rather than relying on the idiosyncratic priorities of faculty members and their highly variable teaching skills, electronic media provide the opportunity to engage learners and facilitate a uniform information base. As discussed in more detail in prior chapters, the potential to create nationally available educational programs is appealing even though individual faculty members may face (short-term) challenges as these become more widely used. That said, electronic media are well suited to facilitate learning across the continuum from subjects that are static to those that are rapidly changing.

The teaching of anatomy is an excellent example. State-of-the-art learning aids have been developed for virtual dissection, rotation of anatomic images, exploration of relationships, and more. Driven by the gaming community, augmented and virtual reality programs further enhance the fidelity. As this technology is increasingly utilized for medical education, sophisticated, affordable learning programs that enhance the teaching of anatomy will continue to be developed. This may well result in a national medical student curriculum for anatomy that incorporates imaging modalities, virtual dissection, and the latest in augmented and virtual reality. Cadaver-based anatomic dissection may no longer be relevant for the majority, or for any medical students.

While it is relatively easy to envision a national or even global technologically enabled approach to learning a subject such as anatomy, it is important to consider the potential of augmented and virtual reality and similar educational tools that would be applicable more broadly to other, perhaps less obvious disciplines and specialties. A consortium of educators, the Internal Medicine Clerkship Directors, has taken a lead in developing IM Essentials [17] to provide a national perspective on foundational material for all medical students on internal medicine clerkships regardless of specialty or career direction. Specialty-specific professional organizations for clerkship directors and GME program directors are often at the forefront of developing standardized resources and learning tools for students and trainees in their clinical disciplines [18–20]. However, utilizing and coordinating these learning resources throughout the medical students' educational journey rarely occurs. Keeping in mind retrieval practice and interleaving principles, how best to further strengthen student learning using the power of technologic enabled educational programs requires thoughtful, collaborative and coordinated developmental efforts.

While achieving a national consensus on discipline-based content and skills is important, the greater benefit is the opportunity to fully utilize the advantages technology provides, and not simply publish a standardized print or electronic textbook under an organization's brand. Building platforms that couple the fidelity available through technology with embedded testing, formative feedback, and analytics to identify problematic areas for the learner is an opportunity to provide educationally powerful clinical scenarios of a multitude of disease presentations, clinical procedures and an array of medical applications. The partnership of medical educators and video game developers has exciting potential, embedding appropriate learning principles within the power of technology.

References

[1] Glover JA. The "testing" phenomenon: not gone but nearly forgotten. J Educ Psychol 1989;81(3):392–9.
[2] Van Gog T, Sweller J. Not new, but not nearly forgotten: the testing effect decreases or even disappears as the complexity of learning materials increases. Educ Psychol Rev 2015;27:247–64.
[3] Taylor K, Rohrer D. The effects of interleaved practice. Appl Cogn Psychol 2010;24:837–48.
[4] Brauer DG, Ferguson KJ. The integrated curriculum in medical education: AMEE guide no. 96. Med Teach 2015;37(4):312–22. https://doi.org/10.3109/0142159X.2014.970998.

[5] http://www.growthengineering.co.uk/what-is-the-forgetting-curve/. (Accessed March 7, 2018).

[6] Kerfoot BP, DeWolf WC, Masser BA, Church PA, Federman DD. Spaced education improves the retention of clinical knowledge by medical students: a randomized controlled trial. Med Educ 2007;41:23–31.

[7] Kerfoot BP, Fu Y, Baker H, Connelly D, Ritchey ML, Genega EM. Online spaced education generates transfer and improves long-term retention of diagnostic skills: a randomized controlled trial. J Am Coll Surg 2010;211(3):331–7.

[8] Tolsgaard MG, Kulasegaram KM, Ringsted C. Practical trials in medical education: linking theory, practice and decision making. Med Educ 2017;51:22–30.

[9] Taylor DCM, Hamdy H. Adult learning theories: implications for learning and teaching in medical education: AMEE guide no. 83. Med Teach 2013;35(11):e1561–72. https://doi.org/10.3109/0142159X.2013.828153.

[10] Gibbs T, Durning S, van der Vleuten C. Theories in medical education: towards creating a union between educational practice and research traditions. Med Teach 2011;33(3):183–7. https://doi.org/10.3109/0142159X.2011.551680.

[11] Gaunt T, Loffman C. When I say...threshold concepts. Med Educ 2018;52:789–90.

[12] Bhat C, Burm S, Mohan T, Chahine S, Goldszmict M. What trainees grapple with: a study of threshold concepts on the medicine ward. Med Educ 2018;52:620–31.

[13] Neve H, Wearn A, Collett T. What are threshold concepts and how can they inform medical education. Med Teach 2016;38(8):850–3.

[14] Sweller J, van Merrienboer JJG, Pass F. Cognitive architecture and instructional design: 20 years later. Educ Psychol Rev 2019. https://doi.org/10.1007/s10648-019-09465-5. Published online: January 22.

[15] Korbach A, Brunken R, Park B. Differentiating different types of cognitive load: a comparison of different measures. Educ Psychol Rev 2018;30:503–29.

[16] Curtis JR, Back AL, Ford DW, Downey L, Shannon SE, Doorenbos AZ. Effect of communication skills training for residents and nurse practitioners on quality of communication with patients with serious illness: a randomized trial. JAMA 2013;310(21):2271–81.

[17] https://ime.acponline.org. (Accessed March 7, 2018).

[18] Clinical Competency Committee Collaborative Learning Community. https://www.im.org/p/cm/ld/fid=1161.

[19] Association for Surgical Education. https://www.surgicaleducation.com.

[20] Association of Pediatric Program Directors. https://www.appd.org/home/index.cfm.

Additional resources

- National Academies of Sciences, Engineering, and Medicine. How people learn II: learners, contexts, and cultures. Washington, DC: The National Academies Press; 2018. https://doi.org/10.17226/24783.
- Lang C, Siemens G, Wise A, Gasevic D, editors. Handbook of learning analytics. 1st ed. Ann Arbor, MI: Society for Learning Analytics Research; 2017.
- Brown PC, Roediger III HL, McDaniel MA. Make it stick: the science of successful learning. Cambridge, MA; London: The Belknap Press of Harvard University Press; 2014.
- SecondLook Histology Series. http://secondlook.med.umich.edu/histology.
- Essential Anatomy 5. https://3d4medical.com/apps/essential-anatomy-5.
- Schuwirth LWT, van der Vleuten CPM. General overview of the theories used in assessment: AMEE guide no. 57. Med Teach 2011;33(10):783–97. https://doi.org/10.3109/0142159X.2011.611022.
- Dolmans DHJM, Tigelaar D. Building bridges between theory and practice in medical education using a design-based research approach: AMEE guide no. 60. Med Teach 2012;34(1):1–10. https://doi.org/10.3109/0142159X.2011.595437.

- Masters K, Ellaway RH, Topps D, Archibald D, Hogue RJ. Mobile technologies in medical education: AMEE guide no. 105. Med Teach 2016;38(6):537–49. https://doi.org/10.3109/0142159X.2016.1141190.
- Yardley S, Teunissen PW, Dornan T. Experiential learning: AMEE guide no. 63. Med Teach 2012;34(2):e102–15. https://doi.org/10.3109/0142159X.2012.650741.
- National Academies of Sciences, Engineering, and Medicine. Improving health professional education and practice through technology: proceedings of a workshop. Washington, DC: The National Academies Press; 2018. https://doi.org/10.17226/25072.

Reflections and concluding thoughts

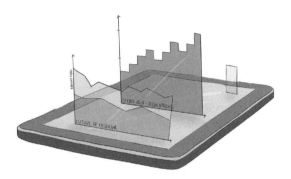

Looking into the future might be considered a perilous or even fool-hardy task. Attempting to predict how the future will look for medicine and then using those predictions to guide the future of medical education might be viewed by the naysayers as at best unrealistic and at worst irresponsible. I say it is necessary and it would be irresponsible not to do so.

What I have put forth is exciting but also anxiety provoking, and in some ways threatening, to those who are in the midst of their careers. Change, especially disruptive rather than incremental change, is usually unwelcome by those who will be most affected. Nevertheless, clinical practice and the clinical care system will assuredly change and likely be subject to disruption in the next few decades. Physicians' practices by midcentury will be markedly different when compared to the early part of this century.

While unforeseen developments will undoubtedly affect clinical medicine, many of the currents of change are already apparent. We just need to look and to be aware. Building on the thoughts expressed by those who I had the privilege to interview, in this book I provide to all who are interested in the education and well-being of our future physicians, as well as to other clinical professionals, a broad overview of these currents of

Implementing Biomedical Innovations into Health, Education, and Practice
https://doi.org/10.1016/B978-0-12-819620-5.00020-5

change, along with examples from the literature that substantiate their validity. Indeed, one of my concerns is that the picture I have presented is too conservative. The reality that will be experienced by many of our students in their midcentury practices might well be even more radically different.

That said, timing is everything. Failure will result if the envisioned changes and their drivers are too far over the horizon and therefore dismissed by those who will be affected. Or, if necessary changes are made too late we will not uphold the implicit social contract we have long enjoyed as a profession.

While some premier academic medical centers might transform themselves to thrive in the new realities, many will not avoid the fate experienced by former leaders in other industries. Whether due to hubris, catastrophic leadership decisions, or the inability to foresee their own demise, some premier academic medical centers will be unable to change and will thus become irrelevant.

Educational time

Time is the most precious commodity any individual has. This is especially true for students and residents. It is incumbent upon medical educators to maximize the educational value of the time we have with our learners. Rather than arbitrarily reducing or lengthening the time spent in training, the focus should be on realistically determining the time needed to acquire the necessary knowledge and learn the skills to function throughout their professional careers. Eliminating irrelevant or marginally useful premedical requirements, medical student courses (and less than useful learning experiences), and resident training requirements that are not consonant with this vision for the future of clinical medicine is challenging but it is urgent work that must be done. It is completely backwards to choose an arbitrary timeframe and then force learning into that timeframe.

Disrupting the status quo

I recognize that what I have recommended will seriously affect the business model of higher education as well as the business plans of academic medical centers. Administrative leaders have faculty and legacy programs that are often supported by tuition. Health systems have come to depend on residents and fellows for patient care and will have to develop alternatives when clinical care responsibilities are not consonant with educational priorities. As such, I fully expect that there will be resistance to some and likely many of the thoughts put forth in this book. Resistance will only grow as the financial foundations of higher education and hospital systems erode. That said, the medical community can either join together to thoughtfully steer the great "medical education and training ship" or allow it to become a rudderless vessel buffeted by external forces. Do we collectively have the will and the determination to embrace and to lead thoughtful change? If we do not, our students and trainees, our future patients, the medical profession, and society as a whole will be the losers.

Beware of "breakthroughs"

While I was in medical school, a senior faculty member advised me to always be skeptical of "breakthrough discoveries" reported by investigators who had the disease they were studying. The lesson was that their personal stake in the research outcomes might bias their judgment and their interpretation of the data. Now I would modify that advice to be skeptical of "breakthrough discoveries" touted by investigators, companies, and institutions that have significant reputational and monetary gains at stake. While it is unquestioned that public-private collaboration is necessary to move true discoveries forward, the conflicts inherent in the growing medical industrial complex require medical professionals to be diligent in their assessment of the efficacy of a product prior to clinical use.

In addition, the growing number of journals, investigators and their institutions issuing attention-grabbing, self-promoting, press releases regarding preliminary discoveries that could potentially be important and falsely raise the hopes of desperate patients and their families is antithetical to the medical profession. Measured and realistic press releases may not generate social media buzz but that is the responsible approach when a clinical application is realistically years away.

As discussed in earlier chapters, true medical breakthroughs must be considered as we shape the future of medical education. But academic medical centers need be cautious and thoughtful as they consider the short-term and long-term consequences of raising expectations that are not readily translated into clinical applications.

Danger of the siren song

The human interaction has been an essential part of the foundation of being a physician that has endured for millennia. Medical educators must resist the temptation to focus on the latest scientific and technologic "advances" to the detriment of what is constant and immutable. Conveying the passion for discovery and progress is vital. However, it must be in balance with the knowledge (and concomitant skills) of the human interaction that will remain <u>central</u> to the practice of medicine. Importantly, medical educators must recognize the core precepts of physicianhood as the settings and accouterments will surely change. Care in the home, asynchronous communication, and the stethoscope becoming as irrelevant as the head mirror, the list of changes is extensive but the focus must be on the constants not the latest technologic "advance."

Diagnosis

Naming is not explaining. As more is understood about the development of disease a diagnostic label should denote the mechanistic basis for the disease process. Already the medical professions continued use of "historical" diagnostic designations results in confusion in the lay public. To many people outside of medicine the differences between type 1 and type 2 diabetes are hazy. Adding to the complexity, it now appears that there are multiple subsets within type 2 diabetes that have different attendant morbidities and likely are different diseases when pathophysiologic mechanisms are considered. Potentially, there are 80 or more different "diseases" that are all lumped under the moniker diabetes mellitus.

But it is not only the laity who suffer from this diagnostic labeling imprecision, so does the profession. Given the pace of scientific discovery it is likely that an ever increasing number of diagnoses will be further subclassified as the underlying pathophysiologic disease mechanism is elucidated. Already we have renal tubular acidosis (RTA) type 1, 2, 3 (may actually be a combination of type 1 and 2 and not a distinct entity) and type 4; multiple endocrine neoplasia (MEN) type 1, 2A, 2B, familial medullary thyroid cancer and type 4; and the list goes on. Rather than perpetuating an ad hoc system built on historical diagnostic labels it is time to take up the challenge issued by the National Academies in their charge to the Committee on a Framework for Developing a New Taxonomy of Disease to "explore the feasibility and need, and develop a potential framework, for creating a 'New Taxonomy' of human diseases based on molecular biology." This will be disruptive!

However, recognizing the power implicit in a diagnosis we should, to the best of our current scientific understanding, accurately label diseases, especially when different therapeutic approaches are indicated.

Fiscal realities

Clinical care and biomedical research has enjoyed unprecedented financial support for the last 50+ years. To use an economic term, it could be characterized as a "bubble." Academic medical centers have profited greatly. However, it is unlikely that this trajectory will continue and faculty and institutional leaders should not be surprised when financial support for our clinical and research activities ceases to expand and potentially contracts.

Technologic innovations will enable diagnostic and therapeutic interventions to be asynchronous in time and geography. The expectation that individuals will seek care through hospitals and clinics will end. Rather, the expectation will be that the diagnostic and therapeutic tools will be made available to patients, on demand, in their homes and in their communities. Scientific discovery and technologic advances will synergistically disrupt the practice of clinical medicine and the prevailing financial business model. The institutions that risk falling behind are those that invest their resources in solutions that are predicated on continuation of the status quo of financial support.

As the educational mission is less profitable and therefore often seen as less important, institutional priorities will become evident. We will see whether educating future generations of physicians is truly core to the existence of an "academic medical center" or is education simply verbiage in the mission statement.

Aligning rhetoric and reality

For decades, we have obfuscated the professional roles and focus of our clinical institutions. We talk about health while investing in ever more sophisticated and expensive approaches to caring for the most critically ill. Caring for the sick is a noble calling deserving of investment and we will continue to need well-trained physicians to care for the sick. I am advocating for clarity as to the clinical focus for physicians.

Potentially this will mean broadening their roles as the opportunity to intervene in high risk and presymptomatic individuals, often through behavioral and environmental interventions rather than with a pill or procedure, becomes a reality. Since many of these interventions are behavioral and do not involve a medicine or a procedure, it is likely that the monetary incentive for clinical care systems and physicians will be minimal.

What roles will we prepare our students and residents to fulfill? What role will we proffer to our patients and society? Will the medical profession and our clinical systems embrace "healthcare" or cede those roles to others?

Restoring meaning and purpose

"I'm worried that if what physicians do in their profession doesn't change, then we're going to stop attracting the people that are the brightest into our field.

IV. Implications for educators

We need to stop telling students that you are going to put your innovative life on hold to get through a trade school. And then when you enter the trade, you can get back to the interesting things you were doing. There's no reason they cannot be synergistic and in fact if defining the core as being a leader, a team member, an impactful person that thinks about health outcomes and tries to make improvement, why not do that during medical school?

We have to resurrect meaning and purpose. In 30 years we're either in one of two scenarios, where we have just allowed things to happen to us and we have been reduced to administrators who kind of just go through the motions or we really embrace a fundamental change in our profession."

Rajesh Mangrulkar

Regardless of scientific and technologic advances, the need for physicians to be "present" and partner with individuals at major transitions in their lifecycle will remain. Medicine must return to its roots as a caring profession. Making a difference in patients' lives is central to restoring meaning and purpose to the medical profession. Technologic advances will free physicians to attend to the human to human interaction, whether in person or asynchronously, reversing the trend toward physicians as well-trained and sophisticated technicians. We must teach our future physicians how to be doctors, and not super-technicians. Unfortunately, some of the faculty physicians who are involved with students also need to (re)learn this, representing a challenge with respect to who will actually do the teaching.

The current crisis of "burnout" within the medical profession has led to an array of interventions with the purported goal of increasing the resilience of individual physicians. However, this phenomenon of burnout has less to do with the individual physicians than the institutional culture. The harsh reality in too many of our clinical systems is that thinking is not respected, honored, or rewarded. The tyranny of efficiency, scorecards measuring metrics and the adherence to guidelines that may or may not be in the best interest of one's patients, reducing the intellectual and professional contributions of the physician to their relative value unit (RVU) production and the increasing emphasis on billings (often above all else) is antithetical to what motivated most physicians to enter medicine. It is incumbent upon our academic and clinical leaders to rectify the negative consequences of what were often well-intentioned actions.

As burnout is also prevalent in residents and medical students, we must identify institutional factors that lead to the erosion of the meaning and purpose that is central to the profession at the local as well as national level. The pernicious effects of high-stakes examinations such as the MCAT and USMLE must be reversed. We tell students aspiring to enter medical school to pursue a well-rounded undergraduate education but then counsel them to focus on the subjects that will be included in the MCAT as that is crucial in admission committee decisions. Indeed, college is too often distilled into the formulaic GPA + MCAT = admission to medical school thus limiting their opportunities to explore other courses and experiences.

Medical schools are reforming their teaching content and their educational processes to reflect what is important in their students' professional development and future practices. Yet it is no secret that to successfully compete for desired residency positions it is critical to do well on USMLE step 1. This is despite studies that have shown the minimal predictive value of USMLE step 1 on performance as a physician. Not surprisingly, this "parallel curriculum" of preparing for the USMLE step 1 exam by pouring through commercial study aids rather than through the coursework offered by their medical school results in cynicism and "burnout."

We establish regulations to govern residency programs, such as duty hours, that may not allow the trainee to continue participating in a clinical event or a procedure that would be very educational. We flaunt the regulations and yet require the trainees to provide patient care that exceeds the time limits and has but marginal educational value. These disconnects must be eliminated and it must be done soon, before it becomes too late to right the ship.

Fortunately or unfortunately, it is our faculty who are responsible for test development and the decisions to accept or reject these metrics when making admission decisions. While there have been efforts to modify the standard examinations by changing the content and question format, it is time for our academic faculty to stop deferring to USMLE scores for residency match decisions. I realize the magnitude of the recommendation I am making. Medical school admissions committees process thousands of applications and medical students are routinely applying to 30–50 or even more residency programs requiring that program directors develop a means to prioritize the hundreds or even thousands of applications received. We must engage in the work of developing alternative and more reality-based methods to assess progress toward agreed-upon, well-defined competencies that will replace these high-stakes, but essentially pointless examinations.

Simultaneously, medical faculty, in collaboration with undergraduate colleagues, need to develop nationally agreed upon competencies and metrics required for medical school in order to ensure that incoming medical students are appropriately prepared. This is all within the purview of the Academy.

A call to action

Considering the trajectory of many individuals who are entering medicine, the students who are matriculating into our colleges and universities this fall will enter practice 15+ years hence. My calculation is as follows:

- Even though the traditional 4 years of college is increasingly 5 years. (We will use 4 years.)

- Often there are 1–2 additional years of education, work, or other experiences before matriculation into medical school. (We will use 1 year.)
- As many medical students pursue research rotations, acquire additional degrees, and take leaves of absence, what used to be 4 years now stretches into five or more years before graduation from medical school. (We will use 5 years.)
- Residency programs range from 3 to 7 years in length (We will use 4 years.)
- And residencies are often followed by further subspecialty or fellowship training that ranges from 1 to 3 years in length. (We will use 1 year.)

This 15 year pipeline, while variable from person-to-person and discipline to discipline, is a stark reminder of the importance of paying attention to not the topic du jour but to that which will provide a foundation for lifelong practice and to the new content and skill areas that must be developed in medical students and residents.

Developing the collaborations required to create the educational experiences beyond traditional medical boundaries, those that allow our students to acquire the knowledge and skills they will need in their professional careers, is an exigent need. Faculty and students in other health professions as well as in engineering, public health, theater and performing arts, business and public policy are among those who would enhance and enrich the medical learning environment. Concomitantly, the unpleasant task of eliminating outmoded educational processes and content from our medical curricula cannot continue to be avoided.

We are currently doing an excellent job of preparing medical students and residents to be physicians in 2020. We are not preparing them to be physicians in 2050. We are creating physicians who are out of sync with the needs of society, who will be able to fill roles that will no longer exist and will not be prepared for the realities they will encounter. We are not instilling in our medical students and residents that technology and advances in computation will have a profound impact on many current disciplines, that many specialties will either be radically changed in their clinical focus or will disappear entirely. Is it educationally or ethically justifiable to continue programs as if the current status will continue ad infinitum when we know it will not?

The scientific and technologic advances we are witnessing, and predicting, have the potential for great good. However, they also have the potential for further distancing physicians from their patients, may diminish patient agency and portend major ethical issues. A fundamental question is whether physicians will embrace the role of "healer" and reemphasize the centrality of the human to human interaction or devolve into ever more sophisticated technicians, ceding the ethical and social responsibilities that have long been part of the ethos of physicianhood. What are we willing to give up in terms of personal contact? What will the patients be willing to give up if technology takes over the human factors? What will be the attributes that admission committees will look for when reviewing medical school applicants? What will be emphasized in formal curriculum and learning experiences? Who will be lauded as the role models for students and residents—the technically sophisticated? The scientifically accomplished? The doctors' Doctor?

Medical education and academic medicine leaders are in control of their destiny. The vision of the future of medical practice for which we are preparing our students and residents must drive our educational programs. If this leadership is lacking, our students and trainees, their future patients and society at large will be the ones paying the price. We must not allow that to happen (Fig. 20.1).

Gone fishing!

It's not hubris, I do "walk on water" some months of the year!

FIG. 20.1   Author enjoying one his favorite outdoor activities—shown here fly fishing on the North Branch of the AuSable River and ice fishing on Deppmann Lake.

Interviewee bios, about the artist

I am grateful to the many distinguished individuals who freely shared their perspectives as to what the future will bring to medical education, to clinical practice and to society at large. As is evidenced from their brief biographical descriptions, the individuals I was privileged to interview represent an array of backgrounds and experiences. The one commonality they share is excellence in their chosen fields.

A multitude of different perspectives emerged, not surprisingly, sometimes diametrically opposed. Some perspectives were commonly cited, others infrequently. From these multiple inputs I have attempted to weave a word tapestry that depicts the future. What is set forth in this book does not represent the views of any one individual and I assume full responsibility for the decision-making that went into crafting what to consider.

All quotes attributed to individual interviewees have been given with permission.

Dan Atkins III, PhD

Daniel E. Atkins III is Emeritus W.K. Kellogg Professor of Information and Professor of Electrical Engineering and Computer Science at the University of Michigan (UM). He did pioneering work on computer architecture, including high-speed arithmetic methods now widely used in modern computers, and application-specific experimental machines. Subsequently he focused on cyber-enabled distributed knowledge communities.

He served as Dean of Engineering, Founding Dean of the School of Information, and Associate VP for Research at UM, as well as the inaugural director of the Office of Cyberinfrastructure at the National Science Foundation (NSF). He chaired the Blue Ribbon Panel on Research Cyberinfrastructure for the NSF that became an international roadmap for initiatives on cyber-enabled research in the digital age. Dr. Atkins received the Nina W. Mathesson Award for contributions to medical informatics and the Paul Evan Peters Award for the use of information resources and services that advance scholarship and intellectual productivity. He is a member of the National Academy of Engineering.

Brian D. Athey, PhD

Brian Athey, the Michael Savageau Professor and founding Chair of the Department of Computational Medicine and Bioinformatics in the University of Michigan Medical School, is an active researcher in "pharmacoepigenomics," a field he helped establish, and is a leader in the field of neuropsychiatric pharmacogenomics. Among the initiatives he has led are the National Library of Medicine (NLM) Visible Human Project, the DARPA Virtual Soldier Project, and the NIH National Center for Integrative Biomedical Informatics (NCIBI).

Dr. Athey is an elected fellow of the American College of Medical Informatics.

Samuel E. Broder, MD, PhD

Samuel Broder has had a distinguished career and major leadership roles in the public and private sectors. He co-developed zidovudine (AZT), didanosine, and zalcitabine which were the first effective drugs licensed for the treatment of AIDS.

From 1989 to 1995 he served as Director of the National Cancer Institute (NCI) and oversaw the development of a number of new therapies for cancer including paclitaxel (Taxol). Subsequently, he became Senior Vice President for Research and Development at the IVAX Corporation, a pharmaceutical company; during the initial sequencing of the human genome race he served as Executive Vice President and Chief Medical Officer of Celera Corporation; then becoming Senior Vice President and Head of INTREXON's (a synthetic biology company) Health Sector.

Dr. Broder's contributions to the public good and medical science have been recognized with receipt of the Arthur S. Flemming Award and the Leopold Griffuel Award. He is a member of the National Academy of Medicine.

Erik W. Brown, PhD

Eric Brown, formerly the Director of Foundational Innovation, IBM Research Healthcare and Life Sciences, led the team responsible for developing advanced cognitive computing technologies for Watson Health.

Francis Collins, MD, PhD

Francis Collins is a physician-geneticist, a pioneer in the discovery of disease genes and Director of the National Institutes of Health (NIH). As the Director of the largest supporter of biomedical research in the world, he oversees research spanning the spectrum from basic to clinical.

In recognition of his contributions to biomedical science Dr. Collins was awarded the Presidential Medal of Freedom in November 2007 and received the National Medal of Science in 2009. He is an elected member of the National Academy of Medicine, National Academy of Sciences and the American Academy of Arts and Sciences National Academy of Sciences.

Lawrence Corey, MD

Lawrence Corey is an internationally recognized expert in virology, immunology and vaccine development. His research led to antivirals for herpes viruses, HIV and hepatitis; the development of experimental vaccines, and new methods for viral infection diagnosis and monitoring. He directed the AIDS Clinical Trials Group and was principal investigator of the HIV Vaccine Trials Network.

He is President and Director Emeritus, Fred Hutchinson Cancer Research Center in Seattle, Washington.

Dr. Correy's contributions to the advancement of biomedical science have been recognized with multiple awards including the Pan American Society of Clinical Virology Award, the American Society for STD Research Parran Award, and the Infectious Diseases Society of America Ender's Award. He is an elected member of the National Academy of Medicine and the American Academy of Arts and Sciences.

James W. Curran, MD, MPH

James Curran, Dean of the Rollins School of Public Health at Emory University, was one of the earliest leaders in the Center for Disease Control's (CDC) response to the then mysterious new disease we now recognize as AIDS. For 15 years, beginning in 1981, through his leadership of the task force on HIV/AIDS and subsequently the HIV/AIDS Division, he was involved in every aspect of determining the epidemiology and response to the HIV/AIDS epidemic.

In recognition of his leadership and scientific contributions to our understanding of HIV/AIDS he received many United States Public Health Service Awards including the Surgeon General's Medal of Excellence, the Edward Brandt, Jr., Award from the National Leadership Coalition Against AIDS, and the John Snow Award from the American Public Health Association. He is an elected member of the National Academy of medicine.

James Duderstadt, PhD

James Duderstadt, President Emeritus and University Professor of Science and Engineering at the University of Michigan, is internationally recognized for his expertise in nuclear energy, high-powered lasers, computer simulation, information technology, and policy development.

He has served on or chaired many public and private boards including the National Science Board; the Policy and Global Affairs Division of the National Research Council; numerous committees of the National Academies including the Executive Council of the National Academy of Engineering and the Committee on Science, Engineering, and Public Policy; the National Commission on the Future of Higher Education; and the Nuclear Energy Advisory Committee of the Department of Energy. He co-directed the Glion Colloquium (Switzerland), and was a nonresident Senior Fellow of the Brookings Institution.

Dr. Duderstadt's contributions to science and engineering have been recognized by receipt of the E.O. Lawrence Award for excellence in nuclear research, the Arthur Holly Compton Prize for outstanding teaching, the President's National Medal of Technology for exemplary service to the nation, and the Vannevar Bush Award for lifelong contributions to the welfare of the Nation through public service activities in science, technology, and public policy. He is an elected member of the National Academy of Engineering and the American Academy of Arts and Science.

Victor Dzau, MD

Victor Dzau, President of the National Academy of Medicine of the United States National Academy of Sciences, is an internationally recognized leader in translational research, health innovation, and global health care strategy and delivery. He has made major scientific contributions to our understanding of cardiovascular disease and genetics.

He has chaired and served on numerous committees including the board of directors of the Singapore Health Services, the Advisory Committee to the Director of U.S. National Institutes of Health, the International Review Board of the Canadian Institute for Health Research, the NIH's Cardiovascular Disease Advisory Committee and the Association of Academic Health Centers. Prior to being named President of the National Academy of Medicine Dr. Dzau served as Chancellor for Health Affairs at Duke University and President and CEO of the Duke University Medical Center.

His contributions to science and medicine have been recognized with multiple honors including the Max Delbrück Medal, the Ellis Island Medal of Honor, the Henry Freisen International Prize, and the Singapore Public Service Medal. He is an elected member of the European Academy of Sciences and Arts, the National Academy of Medicine and the American Academy of Arts and Science.

Evan Eichler, PhD

Evan Eichler, Professor of Genome Sciences at the University of Washington School of Medicine, is an internationally recognized expert in genome instability and segmental duplication. Using computational and experimental approaches he studies the role of duplicated regions and structural variation in the human genome in evolution and disease.

In recognition of his scientific merit he has received multiple awards including the Curt Stern Award from the American Society of Human Genetics, the AAAS (American Association for the Advancement of Science) Newcomb Cleveland Prize, and was selected to give the Mendel Lecture ("Gilded Pea" Award): European Society of Human Genetics. He is a fellow of the American Association for the Advancement of Science, an elected member of the National Academy of Sciences, and the National Academy of Medicine.

Arthur G. Erdman, PhD, PE

Arthur G. Erdman, the Richard C. Jordan Professor and Morse Alumni Distinguished Teaching Professor of Mechanical Engineering and founder and Director of the Earl E. Bakken Medical Devices Center at the University of Minnesota, specializes in bioengineering, medical device and product design. He regularly collaborates with faculty from many clinical disciplines, has consulted with over 50 companies in mechanical and product design and is co-inventor on almost 50 U.S. patents and 15 foreign patents.

His engineering and design contributions have been recognized with multiple awards including the American Society of Mechanical Engineers (ASME) Savio L.-Y. Woo Translational Biomechanics Medal, the Tekne Award as the "Innovative Collaboration of the Year," the ASME Machine Design Award and the ASME Outstanding Design Educator Award. He is a Fellow of ASME and a Founding Fellow of the American Institute for Medical and Biological Engineering (AIMBE).

Maxine Frankel

Maxine Frankel is an internationally recognized art collector and philanthropist. Supporting the visual and performing arts she not only introduces the public to many artists through exhibitions of their collections at major museums in the United States and abroad, she seeks to educate the public regarding the contributions of the arts and humanities to society. A special interest is the role of the humanities in professional development of physicians' humanistic qualities.

Sharing her expertise, she has served on the boards of several museums and institutions including the Noguchi Museum and Socrates Sculpture Park, the Sphinx Organization and the Storm King Art Center.

Larry Goodman, MD

Larry Goodman, James A. Campbell, MD Distinguished Service Professor, was CEO of Rush University Medical Center and President of Rush University. Under his leadership, Rush University Medical Center became nationally recognized for community engagement and technological innovation. Committed to reducing inequities in health and longevity in adjacent communities an innovative technology partnership and screening protocol was developed to identify individuals who lack sufficient food, adequate housing and other needs—the social determinants of health—and match them to available resources. On three occasions, this initiative has been recognized by receipt of the American Hospital Association's Equity of Care Award.

Prior leadership positions include Senior Vice President of Medical Affairs and Dean of Rush Medical College. and Medical Director of Cook County Hospital (now the John H. Stroger, Jr. Hospital of Cook County).

Sanjay Gupta, MD

Sanjay Gupta, CNN's chief medical correspondent, is widely recognized for his reporting on national and international health related issues. As a practicing neurosurgeon and associate Chief of Neurosurgery at Grady Hospital, he brings the perspective gained as an international journalist as seen through the eyes of an accomplished physician.

A multiple Emmy Award winner he has hosted his own television programs and appeared on many others in addition to regularly contributing to major print publications. He has authored several best selling books such as *Chasing Life and Cheating Death* and *Monday Mornings.*

Leroy Hood, MD, PhD

Leroy Hood has made seminal scientific contributions in immunology, neurobiology, cancer biology and biotechnology, and is a leader in the development of systems biology. He has developed groundbreaking instruments including automated DNA sequencers, DNA synthesizers, protein sequencers, and peptide synthesizers that have enabled major scientific advances. Committed to a systems approach to health and disease he advocates P4 medicine; care that is predictive, preventive, personalized and participatory.

In recognition of his many scientific contributions he has received over 100 honors including the Albert Lasker Award for Basic Medical Research, the Kyoto Prize, and the National Medal of Science.

He is a member of the American Academy of Arts and Sciences and an elected member of all three National Academies—the National Academy of Science, the National Academy of Engineering, and the National Academy of Medicine.

Patricia Hurn, PhD, RN

Patricia Hurn, Dean of the School of Nursing, University of Michigan, is an internationally recognized researcher on stroke and other neurological conditions. As an academic leader she focuses on collaborative biohealth research models and science education innovation.

Prior to becoming Dean she served as vice chancellor for research and innovation in the University of Texas System.

Her contributions to nursing and science have been recognized with receipt of the Norma J. Shoemaker Award for Excellence in Critical Care Nursing from the Society for Critical Care Medicine, the Thomas Willis Award for Basic Science from the American Stroke Association/American Heart Association and being named a Fellow, American Academy of Nursing.

H.V. Jagadish, PhD

H.V. Jagadish, the Bernard A Galler Collegiate Professor of Electrical Engineering and Computer Science at the University of Michigan, studies big data and data science. Focusing on usability, he is a leader in the integration of biomedical data from multiple sources, its analysis, and its presentation in a format that delivers real insights to non-technical decision-makers.

He also serves as Senior Scientific Director of the National Center for Integrative Biomedical Informatics, National Institutes of Health. Prior to joining the University of Michigan he was Head of the Database Research Department at AT&T Bell Labs.

He is a fellow of the Association for Computing Machinery (ACM), "The First Society in Computing," and the <u>American Association for the Advancement of Science</u>. In recognition of his research contributions he has received multiple awards including the ACM SIGMOD Contributions Award and the David E. Liddle Research Excellence Award.

Michael M.E. Johns, MD

Michael Johns, former Emory University Chancellor and Emory Executive Vice President for Health, is internationally recognized for his administrative leadership and the vision he brings to large, complex academic medical care organizations. Having held multiple national professional leadership positions; been a member of corporate boards, University boards, and many non-profit organization boards; served as an adviser to the White House on health care reform; and served as editor and on editorial boards of major medical journals he is recognized for the breadth and thoughtful perspective he brings to deliberations on topics ranging from health policy to health professions education to national health system reform.

Among the multitude of honors he has received, his contributions were recognized in 2015 by receipt of the Castle Connolly National Physician of the Year Award for Lifetime Achievement.

He is a fellow of the American Academy of Arts and Sciences and an elected member of the National Academy of Medicine.

Gary S. Kaplan, MD, FACP, FACPME, FACPE

Gary Kaplan, the chairman and CEO of the Virginia Mason Health System in Seattle, is widely recognized for his exceptional leadership in health care transformation and reform. He developed the Virginia Mason Production System, being the first to adapt the principles of the Toyota Production System to health care to identify and eliminate waste, improve quality and safety, and control cost. This has resulted in Virginia Mason receiving international recognition for innovation, quality, safety and efficiency.

Dr. Kaplan has held leadership positions on multiple national organizations including the Institute for Healthcare Improvement Board of Directors and the National Patient Safety Foundation Lucian Leape Institute. He has been named multiple times to Modern Healthcare magazine's annual list of the "50 Most Influential Physician Executives and Leaders in Healthcare" and their list of the "100 Most Influential People in Healthcare." He is an elected member of the National Academy of Medicine.

Vikas Kheterpal, MD

Vikas Kheterpal, entrepreneur and a nationally recognized expert in healthcare informatics, focuses on healthcare data exchange and interoperability. Bringing a unique combination of information technology skills and deep healthcare domain expertise he has had leadership roles across the organizational life cycle ranging from start-up to global segment market leader.

Formerly the Global General Manager and Vice President for Clinical Information Systems for GE Healthcare, currently he is the principal of Care Evolution, Inc.

John L. King, PhD

John King, William Warner Bishop Collegiate Professor of Information and former Dean, School of Information, University of Michigan, studies the relationships between technical change and social change, concentrating on information technologies and change in social institutions. He has studied the problem of design and development of sophisticated socio-technical information infrastructures in complex organizational and institutional settings including work on privacy issues.

He has contributed his expertise to multiple national committees including the Computing Community Consortium (CCC) established by the National Science Foundation and the Advisory Committee for the Computer and Information Science and Engineering Directorate.

Joseph C. Kolars, MD

Joseph C. Kolars is the Josiah Macy, Jr. Professor of Health Sciences Education and Senior Associate Dean for Education and Global Initiatives at the University of Michigan Medical School. His professional career has had two major themes. Having served as Internal Medicine Residency Program Director at the University of Michigan and at Mayo Clinic, a focus on medical education and medical education research, and a focus on global health and strengthening education systems in low-resource settings. As a consultant to the Bill and Melinda Gates Foundation he developed the forerunner of the NIH Medical Education Partnership Initiative (MEPI). He serves on the Fogarty International Center Advisory Council and is codirector of the University of Michigan/Peking University of Health Sciences Joint Institute for Translational and Clinical Research. In the mid-90s he lived and established the first western-based health care system in China.

His expertise as an educator has led to multiple institutional awards and at the national level the Association of Program Directors of Internal Medicine Distinguished Medical Educator Award, the American Gastroenterological Association Distinguished Educator Award and the Association of American Medical Colleges (AAMC) Abraham Flexner Award for Distinguished Service to Medical Education.

K. Ranga Rama Krishnan, MB, ChB

K. Ranga Rama Krishnan, CEO of the Rush University System for Health and previously Dean of Rush Medical College, is an innovative academic medicine leader and medical educator. He has led a major reorganization of the Rush Medical College's curriculum incorporating flipped classrooms and extensive simulation and developed Team LEAD (for Learn, Engage, Apply, Develop) while serving as Dean of the Duke-NUS Graduate Medical School Singapore (now Duke-NUS Medical School). Not only has he led educational innovation he is also leading medical care redesign and developing external partnerships. Prior to serving as Dean of the Duke-NUS Graduate Medical School Singapore he was chairman of the Department of Psychiatry at Duke.

In recognition of his contributions to psychiatry and academic leadership he has received numerous awards, including the Distinguished Scientist Award from the American Association for Geriatric Psychiatry; the Edward Strecker Award from the University of Pennsylvania and the C. Charles Burlingame Award for his lifetime achievements in psychiatric research and education. For his service to Singapore he received the Public Service Medal (Friend of Singapore) from the president of Singapore. He is an elected member of the National Academy of Medicine.

H. Clifford Lane, MD

H. Clifford Lane serves as the National Institute of Allergy and Infectious Diseases (NIAID) Deputy Director for Clinical Research and Special Projects.

An internationally recognized scientist, he was among the first investigators to study the immunopathogenic mechanisms of HIV making seminal discoveries that helped establish the field of HIV immunopathogenesis. He pioneered the strategies of immunologically compatible bone marrow transplantation and the adoptive transfer of lymphocytes in treating patients with HIV infection.

He served in the U.S. Public Health Service Commissioned Corps achieving the rank of Rear Admiral and served as Assistant Surgeon General.

In recognitions of the scientific contributions, he won the Chevalier de I'Ordre National du Mali award and was awarded the U.S. Public Health Service Distinguished Service Medal. He is an elected member of the Institute of Medicine.

James L. Madara, MD

James L. Madara, MD, serves as the CEO and executive vice president of the American Medical Association (AMA), the nation's largest physician organization. Through his leadership the AMA has developed a visionary long-term strategic plan that includes developing and piloting the medical school and educational networks of tomorrow, creating new connected approaches to chronic disease, and defining how to develop a nationwide authentic system of clinical care. As an extension of this vision, he serves as chairman of Health2047 Inc., an independent, design-driven innovation firm whose mission is to advance the goal of improving the health of the nation.

Modern Healthcare consistently recognizes Dr. Madara as one of the nation's 50 most influential physician executives and one of the nation's 100 most influential people in health care. In recognition of his scientific contributions as a leading pathologist he received the Davenport Award for lifetime achievement in gastrointestinal disease from the American Physiological Society and the Mentoring Award for lifetime achievement from the American Gastroenterological Society. He is an elected member of the Association of American Physicians.

Rajesh Mangrulkar, MD

Rajesh Mangrulkar, the Marguerite S. Roll Professor of Medical Education, Associate Dean for Medical Student Education and Professor of Internal Medicine and Learning Health Sciences at the University of Michigan, is leading the initiative to transform the Medical School's curriculum into a program that will graduate physician leaders who help drive change in patient care, health care delivery and science. Part of that vision was the development of the University's first Center for Inter-professional Education in collaboration with leaders from other health sciences schools.

James R. Mault, MD, FACS

James Mault, entrepreneur and industry leader in health information technology (HIT) and medical devices. Having trained as a cardiothoracic surgeon specializing in thoracic oncology, transplantation and critical care he combines expertise in biotechnology with expertise in clinical medicine.

Currently serving as President and Chief Medical Officer of CQuentia, a Genomics/PGx Personalized Medicine solution provider, his prior leadership positions include Senior Vice President and Chief Medical Officer of Qualcomm Life, and Director of New Products, Business Development and Clinical Programs for Microsoft's Health Solutions Group.

He has founded companies, is the named inventor of over 80 issued and pending patents for novel health information technology and medical device innovations, and serves on the Board of Directors of several companies and organizations.

Elizabeth Nabel, MD

Elizabeth Nabel, President of Brigham Health and Professor of Medicine at Harvard Medical School, brings a unique perspective to health care based on her experiences as a physician, research scientist, wellness advocate and academic medicine leader. A cardiologist and accomplished physician scientist studying the molecular genetics of cardiovascular disease, she previously served as director of the National Heart, Lung, and Blood Institute. She serves on multiple corporate and philanthropic boards.

In recognition of Dr. Nabel's scientific contributions and leadership she has been awarded the Kober Medal from the Association of American Physicians, the Willem Einthoven Award from Leiden University in the Netherlands, the Amgen-Scientific Achievement Award, and the Eugene Braunwald Academic Mentorship Award from the American Heart Association among many other honors. She is an elected member of the American Academy of the Arts and Sciences, the National Academy of Medicine, and the Association of American Physicians.

Mary D. Naylor, PhD, RN, FAAN

Mary D. Naylor, the Marian S. Ware Professor, University of Pennsylvania School of Nursing and Director, New Courtland Center for Transitions and Health, has advocated for evidence-based changes in health care practices and policies across the globe. She is a leader in the design of care innovations to improve the outcomes of chronically ill older adults. Her work has resulted in the Transitional Care Model.

Dr. Naylor has contributed her expertise to the Medicare Payment Advisory Commission, serves on the RAND Health Board of Advisors and Agency for Healthcare Research and Quality's National Advisory Council. Her contributions have been honored with receipt of the Distinguished Investigator Award, Academy Health. She is an elected member of the National Academy of Medicine.

Steven Nissen, MD

Steven E. Nissen is a cardiologist, researcher and patient advocate. He is Chairman Of Cardiovascular Medicine at the Cleveland Clinic and served as Medical Director of the Cleveland Clinic Cardiovascular Coordinating Center (C5), an organization that directs multi-center clinical trials. In addition to extensive research on cardiac imaging, he has led multiple clinical trials on the treatment of coronary artery disease and has applied his expertise in clinical trials and statistical analyses to studying the scientific integrity of many approved medications.

Dr. Nissen has served on the U.S. Food and Drug Administration CardioRenal Advisory Panel, testified before Congress regarding adverse medication risks and advised members of Congress on health care policy and related issues. He has served on multiple journal editorial boards and received the Gill Heart Institute Award for Outstanding Contributions to Cardiovascular Research.

Douglas S. Paauw, MD, MACP

Douglas Paauw, Rathmann Family Foundation Endowed Chair for Patient-Centered Clinical Education and Professor of Medicine, University of Washington School of Medicine, is a master teacher and clinician. Practicing as a primary care internist he has a special interest in the care of patients with complex sarcoidosis and HIV. His commitment to the education of future physicians and expertise as a teacher and mentor has been recognized with multiple institutional awards and being named University of Washington School of Medicine Teacher Superior in Perpetuity. He received the Robert Glaser Distinguished Teacher Award from the Alpha Omega Alpha (AOA) and Association of American Medical Colleges.

Dr. Paauw has served as Governor of the American College of Physicians (ACP) Washington State Chapter and on the AOA Board of Directors.

Stephen M. Papadopoulos, MD

Stephen Papadopoulos served as Executive Vice President and Chief Medical Officer of Barrow Neurological Institute, the world's largest dedicated neurosurgical center and a leader in neurosurgical training, research, and patient care. With interests in spinal surgery and an emphasis in the area of image guided surgery and surgical informatics, Doctor Papadopoulos is an internationally recognized leader in neurosurgery.

He contributed his expertise as a board member to multiple biomedical companies and organizations including serving as President of the Congress of Neurological Surgeons.

Stephen C. Quay, MD, PhD

Steven Quay, Chief Executive Officer and President of Atossa Genetics Inc., is an accomplished physician-scientist, inventor, author, and serial biotechnology entrepreneur. He has founded six companies, completed two successful initial public offerings, invented six drugs that are FDA or EMEA approved, and has been granted 86 US patents.

Dr. Quay's scientific, administrative and business expertise has led to multiple senior leadership positions, service on scientific advisory boards, and appointments as a director and board member of numerous biotechnology companies.

Paul G. Ramsey, MD

Paul Ramsey, serves as Chief Executive Officer of University of Washington Medicine, Executive Vice President for Medical Affairs and Dean of the School of Medicine at the University of Washington. He has a distinguished record of academic medicine leadership having previously served as Chair of the Department of Internal Medicine and Dean since 1997.

Dr. Ramsey received the John P. Hubbard Award from the National Board of Medical Examiners in recognition of his research contributions in the field of medical education evaluation. He is an elected member of the Association of American Physicians and the National Academy of Medicine of the National Academy of Sciences.

Eric B. Schoomaker, MD, PhD, Lieutenant General, U.S. Army (RET)

Eric Schoomaker, the Director of the Uniformed Service University of the Health Sciences (USU) LEAD program, studies the central importance of leadership education and training for health professionals in an interprofessional, team-based setting. In addition to serving as the U.S. Army's top medical officer—the 42nd U.S. Army Surgeon General and Commanding General of the U.S. Army Medical Command, he served in multiple leadership positions including Commanding General of the U.S. Army Medical Research and Materiel Command and Fort Detrick; commanded Walter Reed Army Medical Center & North Atlantic Regional Medical Command in Washington, DC and Chief of the Army Medical Corps.

Dr. Schoomaker is the recipient of numerous military awards, including those from France and Germany, the Dr. Nathan Davis Award from the American Medical Association for outstanding government service, and the Philipp M. Lippe, MD Award from the American Academy of Pain Medicine for contributions to the social and political aspect of pain medicine.

Edward Schulak, BA (Arch)

Edward Schulak is an architect, entrepreneur, national real-estate developer, inventor, international business leader, and director of early-stage life sciences and pharmaceuticals companies. Following a highly successful career as President of International Airport Centers and then Chairman of Metro International Trade Services, he became intrigued with the dramatic advances in biotechnology and genomics. Embarking on an intensive course of self-education he is now recognized for his expertise and focus on life sciences, developing groundbreaking technologies for genetic sequencing, forensics, bioinformatics and drug development.

Ellen Sheets, MD, MBA

Ellen Sheets, Entrepreneur in Residence with Partners Innovation Fund, has a broad range of experience within academic medicine and the medical diagnostic, device and pharmaceutical industries. Trained as a gynecologic oncologist, she was an associate professor at Brigham and Women's Hospital with an active clinical practice and an NIH funded cervical cancer research program prior to assuming senior leadership positions in a series of biotechnology companies.

Having served in positions including Chief Executive Officer, Chief Medical Officer, and President, Dr. Sheets breadth of responsibilities and experiences ranges from leading biomarker discovery, to product development, to clinical trials, to strategy development, and government affairs.

Eric Topol, MD

Eric Topol, Gary & Mary West Endowed Chair of Innovative Medicine and Executive Vice President Scripps Research, is an internationally recognized physician scientist. Built on a distinguished research and clinical career including the development of medications and critical heart care therapies routinely used in medical practice he is among the leaders envisioning the future of precision cardiovascular medicine combining genomics, sensors and preventive cardiology.

Dr. Topol was one of the Top 10 Most Cited Researchers in Medicine, and his contributions to medicine have been recognized with numerous honors including the Simon Dack Award, American College of Cardiology and the Andreas Gruntzig Award, European Society of Cardiology. He is an elected member of the National Academy of Medicine.

Frederick S. Upton, United States representative Michigan 6th Congressional District

Congressman Fred Upton, first elected to Congress in 1986, served as Chairman of the Committee on Energy and Commerce from 2010 to 2016 with jurisdiction over matters concerning energy, healthcare, the environment, telecommunications, commerce, manufacturing, and trade. A long-standing advocate for a greater emphasis on biomedical research to improve the public health, Representative Upton, along with U.S. Rep. Diana DeGette, D-Colorado, launched the 21st Century Cures initiative. After a 3-year process, this bipartisan effort culminated with President Obama signing the 21st Century Cures Act into law on December 13, 2016.

Jenna Wiens, PhD

Jenna Wiens, Morris Wellman Assistant Professor of Computer Science and Engineering (CSE) at the University of Michigan, leads the Machine Learning for Data-Driven Decisions (MLD3) research group. With a focus on machine learning and healthcare, she is developing the computational methods needed to organize, process, and transform data into actionable knowledge. Her work has applications in modeling disease progression and predicting adverse patient outcomes with an emphasis on developing accurate patient risk stratification approaches to reduce the rate of nosocomial infections among patients admitted to hospitals.

Dr. Wiens noteworthy research discoveries have led to her inclusion in Forbes 2015 30 under 30 in Science and Healthcare; and the 2017 MIT Tech Review's list of 35 Innovators Under 35.

About the artist

Victoria Bornstein is a Chicago-based painter, illustrator and designer. She graduated from the University of Michigan's Penny Stamps School of Art Design in 2018 and has been working as a graphic designer and freelance artist ever since.

Her work can be found on her website: www.victoriabornstein.com. And on Instagram: @a_portrait_a_day_.

Index

Note: Page numbers followed by *f* indicate figures, *t* indicate tables, and *b* indicate boxes.